William Starr Dana Parsons

How to Know the Wild Flowers

A Guide to the Names, Haunts, and Habits of our Common Wild Flowers

William Starr Dana Parsons

How to Know the Wild Flowers
A Guide to the Names, Haunts, and Habits of our Common Wild Flowers

ISBN/EAN: 9783337107628

Printed in Europe, USA, Canada, Australia, Japan

Cover: Foto ©berggeist007 / pixelio.de

More available books at **www.hansebooks.com**

HOW TO KNOW THE WILD FLOWERS

A Guide

TO THE NAMES, HAUNTS, AND HABITS OF OUR COMMON WILD FLOWERS

BY

MRS. WILLIAM STARR DANA

ILLUSTRATED BY

MARION SATTERLEE

"The first conscious thought about wild flowers was to find out their names—the first conscious pleasure—and then I began to see so many that I had not previously noticed. Once you wish to identify them, there is nothing escapes, down to the little white chickweed of the path and the moss of the wall."
—RICHARD JEFFERIES

NEW YORK

CHARLES SCRIBNER'S SONS

1893

CONTENTS

341

ONE of these days some one will give us a hand-book of our wild flowers, by the aid of which we shall all be able to name those we gather in our walks without the trouble of analyzing them. In this book we shall have a list of all our flowers arranged according to color, as white flowers, blue flowers, yellow flowers, pink flowers, etc., with place of growth and time of blooming.

JOHN BURROUGHS.

vi

PREFACE

⁄ THE pleasure of a walk in the woods and fields is enhanced a hundredfold by some little knowledge of the flowers which we meet at every turn. Their names alone serve as a clew to their entire histories, giving us that sense of companionship with our surroundings which is so necessary to the full enjoyment of outdoor life. . But if we have never studied botany it has been no easy matter to learn these names, for we find that the very people who have always lived among the flowers are often ignorant of even their common titles, and frequently increase our eventual confusion by naming them incorrectly. While it is more than probable that any attempt to attain our end by means of some "Key," which positively bristles with technical terms and outlandish titles, has only led us to replace the volume in despair, sighing with Emerson, that these scholars

> Love not the flower they pluck, and know it not,
> And all their botany is Latin names !

So we have ventured to hope that such a book as this will not be altogether unwelcome, and that our readers will find that even a bowing acquaintance with the flowers repays one generously for the effort expended in its achievement. ⸱Such an acquaintance serves to transmute the tedium of a railway journey into the excitement of a tour of discovery. It causes the monotony of a drive through an ordinarily uninteresting country to be forgotten in the diversion of noting the wayside flowers, and counting a hundred different species where formerly less than a dozen would have been detected. It invests each boggy meadow and bit of rocky woodland with almost irresistible charm.

Surely Sir John Lubbock is right in maintaining that "those who love Nature can never be dull," provided that love be expressed by an intelligent interest rather than by a purely sentimental rapture.

Ninety-seven of the one hundred and four plates in this book are from original drawings from nature. Of the remaining seven plates, six (Nos. LXXX., XCIX., CI., XXII., XLII., LXXXI.), and the illustration of the complete flower, in the Explanation of Terms, are adapted with alterations from standard authors, part of the work in the first three plates mentioned being original. Plate IV. has been adapted from " American Medicinal Plants," by kind permission of the author, Dr. C. F. Millspaugh. The reader should always consult the " Flower Descriptions " in order to learn the actual dimensions of the different plants, as it has not always been possible to preserve their relative sizes in the illustrations. The aim in the drawings has been to help the reader to identify the flowers described in the text, and to this end they are presented as simply as possible, with no attempt at artistic arrangement or grouping.

We desire to express our thanks to Miss Harriet Procter, of Cincinnati, for her assistance and encouragement. Acknowledgment of their kind help is also due to Mrs. Seth Doane, of Orleans, Massachusetts, and to Mr. Eugene P. Bicknell, of Riverdale, New York. To Dr. N. L. Britton, of Columbia College, we are indebted for permission to work in the College Herbarium.

NEW YORK, March 15, 1893.

HOW TO USE THE BOOK

MANY difficulties have been encountered in the arrangement of this guide to the flowers. To be really useful such a guide must be of moderate size, easily carried in the woods and fields; yet there are so many flowers, and there is so much to say about them, that we have been obliged to control our selection and descriptions by certain regulations which we hope will commend themselves to the intelligence of our readers and secure their indulgence should any special favorite be conspicuous by its absence.

These regulations may be formulated briefly as follows:

1. Flowers so common as to be generally recognized are omitted, unless some peculiarity or fact in their history entitles them to special mention.

Under this, Buttercups, Wild Roses, Thistles, and others are ruled out.

2. Flowers so inconspicuous as generally to escape notice are usually omitted.

Here Ragweed, Plantain, and others are excluded.

3. Rare flowers and escapes from gardens are usually omitted.

4. Those flowers are chosen for illustration which seem entitled to prominence on account of their beauty, interest, or frequent occurrence.

5. Flowers which have less claim upon the general public than those chosen for illustration and full description, yet which are sufficiently common or conspicuous to arouse occasional curiosity, are necessarily dismissed with as brief a description as seems compatible with their identification.

In parts of New England, New York, New Jersey, Pennsylvania and in the vicinity of Washington, I have been enabled to describe many of our wild flowers from personal observation ; and I have endeavored to increase the usefulness of the book by including as well those comparatively few flowers not found within the range mentioned, but commonly encountered at some point this side of Chicago.

The grouping according to color was suggested by a passage in one of Mr. Burroughs's " Talks about Flowers." It seemed, on careful consideration, to offer an easier identification than any other arrangement. One is constantly asked the name of some " little blue flower," or some " large pink flower," noted by the wayside. While both the size and color of a flower fix themselves in the mind of the casual observer, the color is the more definitely appreciated characteristic of the two and serves far better as a clew to its identification.

When the flowers are brought in from the woods and fields they should be sorted according to color and then traced to their proper places in the various sections. As far as possible the flowers have been arranged according to the seasons' sequence, the spring flowers being placed in the first part of each section, the summer flowers next, and the autumn flowers last.

It has sometimes been difficult to determine the proper position of a flower—blues, purples, and pinks shading so gradually one into another as to cause difference of opinion as to the color of a blossom among the most accurate. So if the object of our search is not found in the first section consulted, we must turn to that other one which seems most likely to include it.

It has seemed best to place in the White section those flowers which are so faintly tinted with other colors as to give a white effect in the mass, or when seen at a distance. Some flowers are so green as to seem almost entitled to a section of their own, but if closely examined the green is found to be so diluted with white as to render them describable by the term *greenish-white*. A white flower veined with pink will also be described in the White section, unless its general effect should be so pink as to entitle it to a position in the Pink section. Such a flower again as the

Painted Cup is placed in the Red section because its floral leaves are so red that probably none but the botanist would appreciate that the actual flowers were yellow. Flowers which fail to suggest any definite color are relegated to the Miscellaneous section.

With the description of each flower is given—

1. Its common English name—if one exists. This may be looked upon as its "nickname," a title attached to it by chance, often endeared to us by long association, the name by which it may be known in one part of the country but not necessarily in another, and about which, consequently, a certain amount of disagreement and confusion often arises.

2. Its scientific name. This compensates for its frequent lack of euphony by its other advantages. It is usually composed of two Latin—or Latinized—words, and is the same in all parts of the world (which fact explains the necessity of its Latin form). Whatever confusion may exist as to a flower's English name, its scientific one is an accomplished fact—except in those rare cases where an undescribed species is encountered—and rarely admits of dispute. The first word of this title indicates the *genus* of the plant. It is a substantive, answering to the last or family name of a person, and shows the relationship of all the plants which bear it. The second word indicates the *species*. It is usually an adjective, which betrays some characteristic of the plant, or it may indicate the part of the country in which it is found, or the person in whose honor it was named.

3. The English title of the larger Family to which the plant belongs. All flowers grouped under this title have in common certain important features which in many cases are too obscure to be easily recognized ; while in others they are quite obvious. One who wishes to identify the flowers with some degree of ease should learn to recognize at sight such Families as present conspicuously characteristic features.

For fuller definitions, explanations, and descriptions than are here given, Gray's text-books and "Manual" should be consulted. After some few flowers have been compared with the partially technical description which prefaces each popular one, little difficulty should be experienced in the use of a botanical key. Many of the measurements and technical descriptions have been based upon Gray's "Manual." It has been thought best to omit any mention of species and varieties not included in the latest edition of that work.

An ordinary magnifying-glass (such as can be bought for seventy-five cents), a sharp penknife, and one or two dissecting-needles will be found useful in the examination of the smaller flowers. The use of a note-book, with jottings as to the date, color, surroundings, etc., of any newly identified flower, is recommended. This habit impresses on the memory easily forgotten but important details. Such a book is also valuable for further reference, both for our own satisfaction when some point which our experience had already determined has been forgotten, and for the settlement of the many questions which are sure to arise among flower-lovers as to the localities in which certain flowers are found, the dates at which they may be expected to appear and disappear, and various other points which even the scientific books sometimes fail to decide.

Some of the flowers described are found along every country highway. It is interesting to note that these wayside flowers may usually be classed among the foreign population. They have been brought to us from Europe in ballast and in loads of grain, and invariably follow in the wake of civilization. Many of our most beautiful native flowers have been crowded out of the hospitable roadside by these aggressive, irresistible, and mischievous invaders ; for Mr. Burroughs points out that nearly all of our troublesome weeds are emigrants from Europe. We must go to the more remote woods and fields if we wish really to know our native plants. Swamps especially offer an eagerly sought asylum to our shy and lovely wild flowers.

LIST OF PLATES

MOST young people find botany a dull study. So it is, as taught from the text-books in the schools ; but study it yourself in the fields and woods, and you will find it a source of perennial delight.

JOHN BURROUGHS.

HOW TO KNOW THE WILD FLOWERS

INTRODUCTORY CHAPTER

UNTIL a comparatively recent period the interest in plants centred largely in the medicinal properties, and sometimes in the supernatural powers, which were attributed to them.

— O who can tell
The hidden power of herbes and might of magick spell ?—

sang Spenser in the " Faerie Queene ; " and to this day the names of many of our wayside plants bear witness, not alone to the healing properties which their owners were supposed to possess, but also to the firm hold which the so-called " doctrine of signatures " had upon the superstitious mind of the public.   In an early work on " The Art of Simpling," by one William Coles, we read as follows : " Yet the mercy of God which is over all his works, maketh Grasse to grow upon the Mountains and Herbes for the use of men, and hath not only stamped upon them a distinct forme, but also given them particular signatures, whereby a man may read, even in legible characters, the use of them." Our hepatica or liver-leaf, owes both its generic and English titles to its leaves, which suggested the form of the organ after which the plant is named, and caused it to be considered " a sovereign remedy against the heat and inflammation of the liver." *

Although his once-renowned system of classification has since been discarded on account of its artificial character, it is probably to Linnæus that the honor is due of having raised the

---

* Lyte.

study of plants to a rank which had never before been accorded it. The Swedish naturalist contrived to inspire his disciples with an enthusiasm and to invest the flowers with a charm and personality which awakened a wide-spread interest in the subject. It is only since his day that the unscientific nature-lover, wandering through those woods and fields where

> — wide around, the marriage of the plants
> Is sweetly solemnized—

has marvelled to find the same laws in vogue in the floral as in the animal world.

To Darwin we owe our knowledge of the significance of color, form, and fragrance in flowers. These subjects have been widely discussed during the last twenty-five years, because of their close connection with the theory of natural selection ; they have also been more or less enlarged upon in modern text-books. Nevertheless, it seems wiser to repeat what is perhaps already known to the reader, and to allude to some of the interesting theories connected with these topics, rather than to incur the risk of obscurity by omitting all explanation of facts and deductions to which it is frequently necessary to refer.

It is agreed that the object of a flower's life is the making of seed, *i.e.*, the continuance of its kind. Consequently its most essential parts are its reproductive organs, the stamens, and the pistil or pistils.

The stamens (p. 11) are the fertilizing organs. These produce the powdery, quickening material called pollen, in little sacs which are borne at the tips of their slender stalks.

The pistil (p. 11) is the seed-bearing organ. The pollen-grains which are deposited on its roughened summit throw out minute tubes which reach the little ovules in the ovary below and quicken them into life.

These two kinds of organs can easily be distinguished in any large, simple, complete flower (p. 10). The pollen of the stamens, and the ovules which line the base of the pistil, can also be detected with the aid of an ordinary magnifying glass.

Now, we have been shown that nature apparently prefers that

the pistil of a flower should not receive its pollen from the stamens in the same flower-cup with itself. Experience teaches that, sometimes, when this happens no seeds result. At other times the seeds appear, but they are less healthy and vigorous than those which are the outcome of *cross-fertilization*—the term used by botanists to describe the quickening of the ovules in one blossom by the pollen from another.

But perhaps we hardly realize the importance of abundant health and vigor in a plant's offspring.

Let us suppose that our eyes are so keen as to enable us to note the different seeds which, during one summer, seek to secure a foothold in some few square inches of the sheltered roadside. The neighboring herb Roberts and jewel-weeds discharge—catapult fashion—several small invaders into the very heart of the little territory. A battalion of silky-tufted seeds from the cracked pods of the milkweed float downward and take lazy possession of the soil, while the heavy rains wash into their immediate vicinity those of the violet from the overhanging bank. The hooked fruit of the stick-tight is finally brushed from the hair of some exasperated animal by the jagged branches of the neighboring thicket and is deposited on the disputed ground, while a bird passing just overhead drops earthward the seed of the partridge berry. The ammunition of the witch-hazel, too, is shot into the midst of this growing colony; to say nothing of a myriad more little squatters that are wafted or washed or dropped or flung upon this one bit of earth, which is thus transformed into a bloodless battle-ground, and which is incapable of yielding nourishment to one-half or one-tenth or even one hundredth of these tiny strugglers for life!

So, to avoid diminishing the vigor of their progeny by *self-fertilization* (the reverse of cross-fertilization), various species take various precautions. In one species the pistil is so placed that the pollen of the neighboring stamens cannot reach it. In others one of these two organs ripens before the other, with the result that the contact of the pollen with the stigma of the pistil would be ineffectual. Often the stamens and pistils are in different flowers, sometimes on different plants. But these

3

pistils must, if possible, receive the necessary pollen in some way and fulfil their destiny by setting seed. And we have been shown that frequently it is brought to them by insects, occasionally by birds, and that sometimes it is blown to them by the winds.

Ingenious devices are resorted to in order to secure these desirable results. Many flowers make themselves useful to the insect world by secreting somewhere within their dainty cups little glands of honey, or, more properly speaking, nectar, for honey is the result of the bees' work. This nectar is highly prized by the insects and is, in many cases, the only object which attracts them to the flowers, although sometimes the pollen, which Darwin believes to have been the only inducement offered formerly, is sought as well.

But of course this nectar fails to induce visits unless the bee's attention is first attracted to the blossom, and it is tempted to explore the premises ; and we now observe the interesting fact that those flowers which depend upon insect-agency for their pollen, usually advertise their whereabouts by wearing bright colors or by exhaling fragrance. It will also be noticed that a flower sufficiently conspicuous to arrest attention by its appearance alone is rarely fragrant.

When, attracted by either of these significant characteristics,— color or fragrance,—the bee alights upon the blossom, it is sometimes guided to the very spot where the nectar lies hidden by markings of some vivid color. Thrusting its head into the heart of the flower for the purpose of extracting the secreted treasure, it unconsciously strikes the stamens with sufficient force to cause them to powder its body with pollen. Soon it flies away to another plant *of the same kind*, where, in repeating the process just described, it unwittingly brushes some of the pollen from the first blossom upon the pistil of the second, where it helps to make new seeds. Thus these busy bees which hum so restlessly through the long summer days are working better than they know and are accomplishing more important feats than the mere honey-making which we usually associate with their ceaseless activity.

4

Those flowers which are dependent upon night-flying in-sects for their pollen, contrive to make themselves noticeable by wearing white or pale yellow,—red, blue, and pink being with difficulty detected in the darkness. They, too, frequently in-dicate their presence by exhaling perfume, which in many cases increases in intensity as the night falls, and a clue to their whereabouts becomes momentarily more necessary. This fact partially accounts for the large proportion of fragrant white flowers. Darwin found that the proportion of sweet-scented white flowers to sweet-scented red ones was 14.6 per cent. of white to 8.2 of red.

We notice also that some of these night-fertilized flowers close during the day, thus insuring themselves against the visits of insects which might rob them of their nectar or pollen, and yet be unfitted by the shape of their bodies to accomplish their fertilization. On the other hand, many blossoms which are dependent upon the sun-loving bees close at night, securing the same advantage.

Then there are flowers which close in the shade, others at the approach of a storm, thus protecting their pollen and nectar from the dissolving rain ; others at the same time every day. Linnæus invented a famous " flower-clock," which indicated the hours of the day by the closing of different flowers. This habit of closing has been called the " sleep of flowers."

There is one far from pleasing class of flowers which entices insect-visitors,—not by attractive colors and alluring fragrance —but " by deceiving flies through their resemblance to putrid meat—imitating the lurid appearance as well as the noisome smell of carrion."* Our common carrion flower, which covers the thickets so profusely in early summer that Thoreau com-plained that every bush and copse near the river emitted an odor which led one to imagine that all the dead dogs in the neighborhood had drifted to its shore, is probably an example of this class, without lurid color, but certainly with a suf-ficiently noisome smell ! Yet this foul odor seems to answer the plant's purpose as well as their delicious aroma does that of

* Grant Allen.

more refined blossoms, if the numberless small flies which it manages to attract are fitted to successfully transmit its pollen.

Certain flowers are obviously adapted to the visits of insects by their irregular forms. The fringed or otherwise conspicuous lip and long nectar-bearing spur of many orchids point to their probable dependence upon insect-agency for perpetuation ; while the papilionaceous blossoms of the Pulse family also betray interesting adaptations for cross-fertilization by the same means. Indeed it is believed that irregularity of form is rarely conspicuous in a blossom that is not visited by insects.

The position of a nodding flower, like the harebell, protects its pollen and nectar from the rain and dew ; while the hairs in the throat of many blossoms answer the same purpose and exclude useless insects as well.

Another class of flowers which calls for special mention is that which is dependent upon the wind for its pollen. It is interesting to observe that this group expends little effort in useless adornment. "The wind bloweth where it listeth" and takes no note of form or color. So here we find those

Wan flowers without a name,

which, unheeded, line the way-side. The common plantain of the country dooryard, from whose long tremulous stamens the light, dry pollen is easily blown, is a familiar example of this usually ignored class. Darwin first observed, that "when a flower is fertilized by the wind it never has a gayly colored corolla." Fragrance and nectar as well are usually denied these sombre blossoms. Such is the occasional economy of that at times most reckless of all spendthrifts—nature !

Some plants—certain violets and the jewel-weeds among others—bear small inconspicuous blossoms which depend upon no outside agency for fertilization. These never open, thus effectually guarding their pollen from the possibility of being blown away by the wind, dissolved by the rain, or stolen by insects. They are called *cleistogamous* flowers.

Nature's clever devices for securing a wide dispersion of seeds have been already hinted at. One is tempted to dwell at length upon the ingenious mechanism of the elastically bursting capsules of one species, and the deft adjustment of the silky sails which waft the seeds of others ; on the barbed fruits which have pressed the most unwilling into their prickly service, and the bright berries which so temptingly invite the hungry winter birds to peck at them till their precious contents are released, or to devour them, digesting only the pulpy covering and allowing the seeds to escape uninjured into the earth at some conveniently remote spot.

Then one would like to pause long enough to note the slow movements of the climbing plants and the uncanny ways of the insect-devourers. At our very feet lie wonders for whose elucidation a lifetime would be far too short. Yet if we study for ourselves the mysteries of the flowers, and, when daunted, seek their interpretation in those devoted students who have made this task part of their life-work, we may hope finally to attain at least a partial insight into those charmed lives which find

> —tongues in trees, books in the running brooks,
> Sermons in stones, and good in everything.

EXPLANATION OF TERMS

THE comprehension of the flower descriptions and of the opening chapters will be facilitated by the reading of the following explanation of terms. For words or expressions other than those which are included in this section, the Index of Technical Terms at the end of the book should be consulted.

The **Root** of a plant is the part which grows downward into the ground and absorbs nourishment from the soil. True roots bear nothing besides root-branches or **rootlets.**

" The **Stem** is the axis of the plant, the part which bears all the other organs." (Gray.)

A **Rootstock** is a creeping stem which grows beneath the surface of the earth. (See Blood-root and Solomon's Seal. Pls. I. and X.)

A **Tuber** is a thickened end of a rootstock, bearing buds, —" eyes,"—on its sides. The common Potato is a familiar example of a tuber, being a portion of the stem of the potato plant.

A **Corm** is a short, thick, fleshy underground stem which sends off roots from its lower face. (See Jack in the Pulpit, Pl. CIV.)

A **Bulb** is an underground stem, the main body of which consists of thickened scales, which are in reality leaves or leaf bases, as in the onion.

A **Simple Stem** is one which does not branch.

A **Stemless** plant is one which bears no obvious stem, but only leaves and flower-stalks, as in the Common Blue Violet and Liver-leaf (Pl. LXXXIV.).

A **Scape** is the leafless flower-stalk of a stemless plant. (See Liver-leaf (Pl. LXXXIV.).

An **Entire Leaf** is one the edge of which is not cut or lobed in any way. (See Rhododendron, Pl. XVI., and Closed Gentian, Pl. C.)

A **Simple Leaf** is one which is not divided into leaflets; its edges may be either lobed or entire. (See Rhododendron, Pl. XVI.; also Fig. 1.)

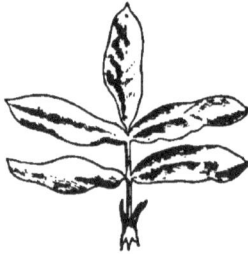

Fig. 1.   Fig. 2.   Fig. 3.

A **Compound Leaf** is one which is divided into leaflets, as in the Wild Rose, Pink Clover, and Travellers' Joy (Pl. XXXI., also Fig. 2).

A **Much-divided Leaf** is one which is several times divided into leaflets (Fig. 3).

The **Axil** of a leaf is the upper angle formed by a leaf or leaf-stalk and the stem.

Flowers which grow from the axils of the leaves are said to be **Axillary**.

A cluster in which the flowers are arranged—each on its own stalk—along the sides of a common stem or stalk is called a **Raceme**. (See Cardinal-flower, Pl. LXXXIII.; Shin-leaf, Pl. XVIII.)

A cluster in which the flower-stalks all spring from apparently the same point, as in the Milkweeds, somewhat suggesting the spreading ribs of an umbrella, is called an **Umbel** (Pl. LXXXI.).

A cluster which is formed of a number of small umbels, all of the stalks of which start from apparently the same point, is called a **Compound Umbel.** (See Wild Carrot, Pl. XXVIII.)

A close, circular flower-cluster, like that of Pink Clover or Dandelion, is called a **Head.** (See Oswego Tea, Pl. LXXXII. ; Sunflower, Pl. LVII.)

A flower-cluster along the lengthened axis of which the flowers are sessile or closely set is called a **Spike.** (See Vervain, Pl. XCII. ; Mullein, Pl. LI.)

A **Spadix** is a fleshy spike or head, with small and often imperfect flowers, as in the Jack-in-the-Pulpit, and Skunk Cabbage (Pls. CII. and CIV., also Fig. 4).

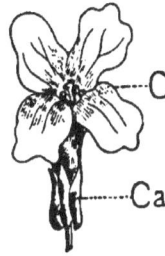

Fig. 4.   Fig. 5.   Fig. 6.

A **Spathe** is the peculiar leaf-like bract which usually envelopes a spadix. (See Jack-in-the-Pulpit and Skunk Cabbage, Pls. CII. and CIV., also Fig. 5.)

A leaf or flower which is set so close in the stem as to show no sign of a separate leaf or flower-stalk, is said to be **Sessile.**

A **Complete Flower** (Fig. 6) is "that part of a plant which subserves the purpose of producing seed, consisting of stamens and pistils, which are the essential organs, and the calyx and corolla, which are the protecting organs." (Gray.)

The green outer flower-cup, or outer set of green leaves, which we notice at the base of many flowers, is the **Calyx** (Fig. 6 Ca). At times this part is brightly colored and may be the most conspicuous feature of the flower.

When the calyx is divided into separate leaves, these leaves are called **Sepals.**

The inner flower-cup or the inner set of leaves is the **Corolla** (Fig. 6, C).

When the corolla is divided into separate leaves, these leaves are called **Petals**.

. We can look upon calyx and corolla as the natural tapestry which protects the delicate organs of the flower, and serves as well, in many cases, to attract the attention of passing insects. In some flowers only one of these two parts is present ; in such a case the single cup or set of floral leaves is generally considered to be the calyx.

The floral leaves may be spoken of collectively as the **Perianth**. This word is used especially in describing members of families where there might be difficulty in deciding as to whether the single set of floral leaves present should be considered calyx or corolla (see Lilies, Pls. XLV. and LXXX.) ; or where the petals and sepals can only be distinguished with difficulty, as with the Orchids.

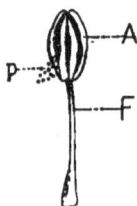 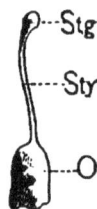

Fig. 7.   Fig. 8.

The **Stamens** (Fig. 7) are the fertilizing organs of the flower. A stamen usually consists of two parts, its **Filament** (F), or stalk, and its **Anther** (A), the little sac at the tip of the filament which produces the dust-like, fertilizing substance called **Pollen** (p.).

The **Pistil** (Fig. 8) is the seed-bearing organ of the flower. When complete it consists of **Ovary** (O), **Style** (Sty.), and **Stigma** (Stg.).

The **Ovary** is the hollow portion at the base of the pistil. It contains the ovules or rudimentary seeds which are quickened into life by the pollen.

The **Style** is the slender tapering stalk above the ovary.

The **Stigma** is usually the tip of the style. The pollen-grains which are deposited upon its moist roughened surface throw out minute tubes which penetrate to the little ovules of the ovary and cause them to ripen into seeds.

A flower which has neither stamens nor pistils is described as **Neutral.**

A flower with only one kind of these organs is termed **Unisexual.**

A **Male** or **Staminate** flower is one with stamens but without pistils.

A **Female** or **Pistillate** flower is one with pistils but without stamens.

The **Fruit** of a plant is the ripened seed-vessel or seed-vessels, including the parts which are intimately connected with it or them.

ALTHOUGH the great majority of plant families can only be distinguished by a combination of characteristics which are too obscure to obtain any general recognition, there are some few instances where these family traits are sufficiently conspicuous to be of great assistance in the ready identification of flowers.

If, for instance, we recognize at sight a papilionaceous blossom and know that such an one only occurs in the Pulse family, we save the time and energy which might otherwise have been expended on the comparison of a newly found blossom of this character with the descriptions of flowers of a different lineage. Consequently it has seemed wise briefly to describe the marked features of such important families as generally admit of easy identification.

Composite Family.—It is fortunate for the amateur botanist that the plant family which usually secures the quickest recognition should also be the largest in the world. The members of the Composite family attract attention in every quarter of the globe, and make themselves evident from early spring till late autumn, but more especially with us during the latter season.

The most noticeable characteristic of the Composites is the crowding of a number of small flowers into a close cluster or *head*, which head is surrounded by an involucre, and has the effect of a single blossom. Although this grouping of small flowers in a head is not peculiar to this tribe, the same thing being found in the clovers, the milkworts, and in various other plants—still a little experience will enable one to distinguish a Composite without any analysis of the separate blossoms which form the head.

13

These heads vary greatly in size and appearance. At times they are large and solitary, as in the dandelion. Again they are small and clustered, as in the yarrow (Pl. XXVIII.).

In some genera they are composed of flowers which are all similar in form and color, as in the dandelion, where all the corollas are *strap-shaped* and yellow; or, as in the common thistle, where they are all *tubular-shaped* and pinkish-purple.

In others they are made up of both kinds of flowers, as in the daisy, where only the yellow central or *disk-flowers* are tubular-shaped, while the white outer or *ray-flowers* are strap-shaped. The flower-heads of the well-known asters and golden rods are composed of both ray and disk-flowers also; but while the ray-flowers of the aster, like those of the daisy, wear a different color from the yellow disk-flowers, both kinds are yellow in the golden rod.

If the dandelion or the chicory (Pl. XCVIII.) is studied as an example of a head which is composed entirely of strap-shaped blossoms; the common thistle or the stick-tight (Pl. LVIII.) as an example of one which is made up of tubular-shaped blossoms; and the daisy or the sunflower (Pl. LVII.) as an example of one which combines ray and disk-flowers—as the strap-shaped and tubular blossoms are called when both are present—there need be little difficulty in the after recognition of a member of this family. The identification of a particular species or even genus will be a less simple matter; the former being a task which has been known to tax the patience of even advanced botanists.

Mr. Grant Allen believes that the Composites largely owe their universal sway to their "co-operative system." He says: "If we look close into the Daisy we see that its centre comprises a whole mass of little yellow bells, each of which consists of corolla, stamens, and pistil. The insect which alights on the head can take his fill in a leisurely way, without moving from his standing-place; and meanwhile he is proving a good ally of the plant by fertilizing one after another of its numerous ovaries. Each tiny bell by itself would prove too inconspicuous to attract much attention from the passing bee; but union is strength for the Daisy as for the State, and the little composites have found their co-operative system answer so well, that late as was their appearance upon the earth they are generally considered at the present day to be the most numerous family both in species and individuals of all flowering plants.'' While those of us who know the country lanes at that season when

—ranks of seeds their witness bear,

feel that much of their omnipresence is due to their unsurpassed facilities for globe-trotting. Our roadsides every autumn are lined with tall golden-rods, whose brown, velvety clusters are composed of masses of tiny seeds whose downy sails are set for their aerial voyage ; with asters, whose myriad flower-heads are transformed into little puff-balls which are awaiting dissolution by the November winds, and with others of the tribe whose hooked seeds win a less ethereal but equally effective transportation.

Parsley Family.—The most familiar representative of the Parsley family is the wild carrot, which so profusely decks the highways throughout the summer with its white, lace-like clusters ; while the meadow parsnip is perhaps the best known of its yellow members.

This family can usually be recognized by the arrangement of its minute flowers in umbels (p. 9), which umbels are again so clustered as to form a compound umbel (Wild Carrot, Pl. XXVIII.) whose radiating stalks suggest the ribs of an umbrella, and give this Order its Latin name of *Umbelliferæ*.

A close examination of the tiny flowers which compose these umbrella-like clusters discovers that each one has five white or yellow petals, five stamens, and a two-styled pistil. Sometimes the calyx shows five minute teeth. The leaves are usually divided into leaflets or segments which are often much toothed or incised.

The Parsleys are largely distinguished from one another by differences in their fruit, which can only be detected with the aid of a microscope. It is hoped, however, that the more common and noticeable species will be recognized by means of

15

descriptions which give their general appearance, season of blooming, and favorite haunts.

Pulse Family.—The Pulse family includes many of our common wood- and field-flowers. The majority of its members are easily distinguished by those irregular, butterfly-shaped blossoms which are described as *papilionaceous*. The sweet pea is a familiar example of such a flower, and a study of its curious structure renders easy the after identification of a papilionaceous blossom, even if it be as small as one of the many which make up the head of the common pink clover.

The calyx of such a flower is of five more or less—and sometimes unequally—united sepals. The corolla consists of five irregular petals, the upper one of which is generally wrapped about the others in bud, while it spreads or turns backward in flower. This petal is called the *standard*. The two side petals are called *wings*. The two lower ones are usually somewhat united and form a sort of pouch which encloses the stamens and style; this is called the *keel*, from a fancied likeness to the prow of an ancient vessel. There are usually ten stamens and one pistil.

These flowers are peculiarly adapted to cross-fertilization through insect agency, although one might imagine the contrary to be the case from the relative positions of stamens and pistil. In the pea-blossom, for example, the hairy portion of the style receives the pollen from the early maturing stamens. The weight of a visiting bee projects the stigma and the pollen-laden style against the insect's body. But it must be observed that in this action the *stigma first brushes against the bee*, while the *pollen-laden style touches him later*, with the result that the bee soon flies to another flower on whose fresh stigma the detached pollen is left, while a new cargo of this valuable material is unconsciously secured, and the same process is indefinitely repeated.

Mint Family.—A member of the Mint family usually exhales an aromatic fragrance which aids us to place it correctly. If to this characteristic is added a square stem, opposite leaves, a two-lipped corolla, four stamens in pairs—two being longer than the others—or two stamens only, and a pistil whose style (two-lobed at the apex) rises from a deeply four-lobed ovary which splits apart in fruit into four little seed-like nutlets, we may feel sure that one of the many Mints is before us.

Sometimes we think we have encountered one of the family because we find the opposite leaves, two-lipped corolla, four stamens, and an ovary that splits into four nutlets in fruit; but unless the ovary was also deeply four-lobed in the flower, the plant is probably a Vervain, a tribe which greatly resembles the Mints. The Figworts, too, might be confused with the Mints did we not always keep in mind the four-lobed ovary.

In this family we find the common catnip and pennyroyal, the pretty ground ivy, and the handsome bee balm (Pl. LXXXII.).

Mustard Family.—The Mustard family is one which is abundantly represented in waste places everywhere by the little shepherd's purse or pickpocket, and along the roadsides by the yellow mustard, wild radish, and winter-cress (Pl. XLII.).

Its members may be recognized by their alternate leaves, their biting harmless juice, and by their white, yellow, or purplish flowers, the structure of which at once betrays the family to which they belong. .

The calyx of these flowers is divided into four sepals. The four petals are placed opposite each other in pairs, their spreading blades forming a cross which gives the Order its Latin name *Cruciferæ*. There are usually six stamens, two of which are inserted lower down than the others. The single pistil becomes in fruit a pod. Many of the Mustards are difficult of identification without a careful examination of their pods and seeds.

Orchis Family.—To the minds of many the term orchid only suggests a tropical air-plant, which is rendered conspicuous either by its beauty or by its unusual and noticeable structure.

This impression is, perhaps, partly due to the rude print in some old text-book which endeared itself to our childish minds by those startling and extravagant illustrations which are responsible for so many shattered illusions in later life; and partly to the various exhibitions of flowers in which only the exotic members of this family are displayed.

Consequently, when the dull clusters of the ragged fringed orchis, or the muddy racemes of the coral-root, or even the slender, graceful spires of the ladies' tresses are brought from the woods or roadside and exhibited as one of so celebrated a tribe, they are usually viewed with scornful incredulity, or, if the authority of the exhibitor be sufficient to conquer disbelief, with unqualified disappointment. The marvellous mechanism which is exhibited by the humblest member of the Orchis family, and which suffices to secure the patient scrutiny and wondering admiration of the scientist, conveys to the uninitiated as little of interest or beauty as would a page of Homer in the original to one without scholarly attainments.

The uprooting of a popular theory must be the work of years, especially when it is impossible to offer as a substitute one which is equally capable of being tersely defined and readily apprehended ; for many seem to hold it a righteous principle to cherish even a delusion till it be replaced by a belief which affords an equal amount of satisfaction. It is simpler to describe an orchid as a tropical air-plant which apes the appearance of an insect and never roots in the ground than it is to master by patient study and observation the various characteristics which so combine in such a plant as to make it finally recognizable and describable. Unfortunately, too, the enumeration of these unsensational details does not appeal to the popular mind, and so fails to win by its accuracy the place already occupied by the incorrect but pleasing conception of an orchid.

For the benefit of those who wish to be able to correctly place these curious and interesting flowers, as brief a description as seems compatible with their recognition is appended.

Leaves.—Alternate, parallel-nerved.

Flowers.—Irregular in form, solitary or clustered, each one subtended by a bract.

Perianth.—Of six divisions in two sets. The three outer divisions are sepals, but they are usually petal-like in appearance. The three inner are petals. By a twist of the ovary what would otherwise be the upper petal is made the lower. This division is termed the *lip ;* it is frequently brightly colored or grotesquely shaped, being at times deeply fringed or furrowed ; it has often a spur-like appendage which secretes nectar ; it is an important feature of the flower and is apparently designed to attract insects for the purpose of securing their aid in the cross-fertilization which is usually necessary for the perpetuation of the different species of this family, all of which give evidence of great modification by means of insect-selection.

In the heart of the flower is the *column ;* this is usually composed of the stamen (of two in the *Cypripediums*), which is confluent with the *style* or thick, fleshy *stigma*. The two *cells* of the *anther* are placed on either side of and somewhat above the stigma ; these cells hold the two pollen masses.

Darwin tells us that the flower of an orchid originally consisted of fifteen different parts, three petals, three sepals, six stamens, and three pistils. He shows traces of all these parts in the modern orchid.

FLOWER DESCRIPTIONS

" A fresh footpath, a fresh flower, a fresh delight "
RICHARD JEFFERIES

21

WHITE

BLOOD-ROOT.

Sanguinaria Canadensis. Poppy Family.

Rootstock.—Thick, charged with a crimson juice. *Scape.*—Naked, one-flowered. *Leaves.*—Rounded, deeply-lobed. *Flower.*—White, terminal. *Calyx.*—Of two sepals falling early. *Corolla.*—Of eight to twelve snow-white petals. *Stamens.*—About twenty-four. *Pistil.*—One, short.

In early April the firm tip of the curled-up leaf of the blood-root pushes through the earth and brown leaves, bearing within its carefully shielded burden—the young erect flower-bud. When the perils of the way are passed and a safe height is reached this pale, deeply lobed leaf resigns its precious charge and gradually unfolds itself; meanwhile the bud slowly swells into a blossom.

Surely no flower of all the year can vie with this in spotless beauty. Its very transitoriness enhances its charm. The snowy petals fall from about their golden centre before one has had time to grow satiated with their perfection. Unless the rocky hillsides and wood-borders are jealously watched it may escape us altogether. One or two warm sunny days will hasten it to maturity, and a few more hours of wind and storm shatter its loveliness.

Care should be taken in picking the flower—if it must be picked—as the red liquid which oozes blood-like from the wounded stem makes a lasting stain. This crimson juice was prized by the Indians for decorating their faces and tomahawks.

SHAD-BUSH. JUNE-BERRY. SERVICE-BERRY.

Amelanchier oblongifolia. Rose Family.

A tall shrub or small tree found in low ground. *Leaves.*—Oblong, acutely pointed, finely toothed, mostly rounded at base. *Flowers.*—White, growing in racemes. *Calyx.*—Five-cleft. *Corolla.*—Of five rather long

PLATE I

BLOOD-ROOT.—*S. Canadensis.*

23

petals. *Stamens.*—Numerous, short. *Pistils.*—With five styles. *Fruit.*—Round, red, berry-like, sweet and edible, ripening in June.

Down in the boggy meadow in early March we can almost fancy that from beneath the solemn purple cowls of the skunk-cabbage brotherhood comes the joyful chorus—

> For lo, the winter is past !—

but we chilly mortals still find the wind so frosty and the woods so unpromising that we return shivering to the fireside and refuse to take up the glad strain till the feathery clusters of the shad-bush droop from the pasture thicket. Then only are we ready to admit that

> The flowers appear upon the earth,
> The time of the singing of birds is come.

Even then, search the woods as we may, we shall hardly find thus early in April another shrub in blossom, unless it be the spice-bush, whose tiny honey-yellow flowers escape all but the careful observer. The shad-bush has been thus named because of its flowering at the season when shad "run;" June-berry, because the shrub's crimson fruit surprises us by gleaming from the copses at the very beginning of summer ; service-berry, because of the use made by the Indians of this fruit, which they gathered in great quantities, and, after much crushing and pounding, utilized in a sort of cake.

WOOD ANEMONE. WIND-FLOWER.

Anemone nemorosa. Crowfoot Family.

Stem.—Slender. *Leaves.*—Divided into delicate leaflets. *Flower.*—Solitary, white, pink, or purplish. *Calyx.*—Of from four to seven petal-like sepals. *Corolla.*—None. *Stamens and Pistils.*—Numerous.

> —Within the woods,
> Whose young and half transparent leaves scarce cast
> A shade, gay circles of anemones
> Danced on their stalks ;

writes Bryant, bringing vividly before us the feathery foliage of the spring woods, and the tremulous beauty of the slender-stemmed anemones. Whittier, too, tells how these

> —wind flowers sway
> Against the throbbing heart of May.

24

PLATE II

RUE ANEMONE.—*A. thalictroides.*   WOOD ANEMONE —*A. nemorosa.*

And in the writings of the ancients as well we could find many allusions to the same flower were we justified in believing that the blossom christened the " wind-shaken," by some poet flower-lover of early Greece, was identical with our modern anemone.

Pliny tells us that the anemone of the classics was so entitled because it opened at the wind's bidding. The Greek tradition claims that it sprang from the passionate tears shed by Venus over the body of the slain Adonis. At one time it was believed that the wind which had passed over a field of anemones was poisoned and that disease followed in its wake. Perhaps because of this superstition the flower was adopted as the emblem of sickness by the Persians. Surely our delicate blossom is far removed from any suggestion of disease or unwholesomeness, seeming instead to hold the very essence of spring and purity in its quivering cup.

RUE ANEMONE.

Anemonella thalictroides. Crowfoot Family.

Stem.—Six to twelve inches high. *Leaves.*—Divided into rounded leaf-lets. *Flowers.*—White or pinkish, clustered. *Calyx.*—Of five to ten petal-like sepals. *Corolla.*—None. *Stamens.*—Numerous. *Pistils.*—Four to fifteen.

The rue anemone seems to linger especially about the spreading roots of old trees. It blossoms with the wood anemone, from which it differs in bearing its flowers in clusters.

STAR-FLOWER.

Trientalis Americana. Primrose Family.

Stem.—Smooth, erect. *Leaves.*—Thin, pointed, whorled at the summit of the stem. *Flowers.*—White, delicate, star-shaped. *Calyx.*—Generally seven-parted. *Corolla.*—Generally seven-parted, flat, spreading.—*Stamens.*—Four or five. *Pistil.*—One.

Finding this delicate flower in the May woods, one is at once reminded of the anemone. The whole effect of plant, leaf, and snow-white blossom is starry and pointed. The frosted tapering petals distinguish it from the rounded blossoms of the wild strawberry, near which it often grows.

PLATE III

STAR FLOWER.
—*T. Americana.*

Fruit.   Flower.

Maianthemum Canadense.

27

Osmelacina ?

Maianthemum Canadense. Lily Family.

Stem.—Three to six inches high, with two or three leaves. *Leaves.*—Lance-shaped to oval, heart-shaped at base. *Flowers.*—White or straw-color, growing in a raceme. *Perianth.*—Four-parted. *Stamens.*—Four. *Pistil.*—One, with a two-lobed stigma. *Fruit.*—A red berry.

It seems unfair that this familiar and pretty little plant should be without any homely English name. Its botanical title signifies " Canada Mayflower," but while it undoubtedly grows in Canada and flowers in May, the name is not a happy one, for it abounds as far south as North Carolina, and is not the first blossom to be entitled " Mayflower."

In late summer the red berries are often found in close proximity to the fruit of the shin-leaf and pipsissewa.

GOLD THREAD.

Coptis trifolia. Crowfoot Family.

Scape.—Slender, three to five inches high. *Leaves.*—Evergreen, shining, divided into three leaflets. *Flowers.*—Small, white, solitary. *Calyx.*—Of five to seven petal-like sepals which fall early. *Corolla.*—Of five to seven club-shaped petals. *Stamens.*—Fifteen to twenty-five. *Pistils.*—Three to seven. *Root.*—Of long, bright yellow fibres.

This little plant abundantly carpets the northern bogs and extends southward over the mountains, its tiny flowers appearing in May. Its bright yellow thread-like roots give it its common name.

PYXIE. FLOWERING-MOSS.

Pyxidanthera barbulata. Order *Diapensiaceæ.*

Stems.—Prostrate and creeping, branching. *Leaves.*—Narrowly lance-shaped, awl-pointed. *Flowers.*—White or pink, small, numerous. *Calyx.*—Of five sepals. *Corolla.*—Five-lobed. *Stamens.*—Five. *Pistil.*—One, with a three-lobed stigma.

In early spring we may look for the white flowers of this moss-like plant in the sandy pine-woods of New Jersey and southward. At Lakewood they appear even before those of the trailing arbutus which grows in the same localities. The generic name is from two Greek words which signify *a small box* and *anther*, and refers to the anthers, which open as if by a lid.

CRINKLE-ROOT. TOOTHWORT. PEPPER-ROOT.

Dentaria diphylla. Mustard Family (p. 17).

Rootstock.—Five to ten inches long, wrinkled, crisp, of a pleasant, pungent taste. *Stem.*—Leafless below, bearing two leaves above. *Leaves.*—Divided into three toothed leaflets. *Flowers.*—White, in a terminal cluster. *Pod.*—Flat and lance-shaped.

The crinkle-root has been valued—not so much on account of its pretty flowers which may be found in the rich May woods —but for its crisp edible root which has lent savor to many a simple luncheon in the cool shadows of the forest.

SPRING-CRESS.

Cardamine rhomboidea. Mustard Family (p. 17).

Rootstock.—Slender, bearing small tubers. *Stem.*—From a tuberous base, upright, slender. *Root-leaves.*—Round and often heart-shaped.— *Stem-leaves.*—The lower rounded, the upper almost lance-shaped. *Flowers.* —White, large. *Pod.*—Flat, lance-shaped, pointed with a slender style tipped with a conspicuous stigma ; smaller than that of the crinkle-root.

The spring-cress grows abundantly in the wet meadows and about the borders of springs. Its large white flowers appear as early as April, lasting until June.

WHITLOW-GRASS.

Draba verna. Mustard Family (p. 17).

Scapes.—One to three inches high. *Leaves.*—All from the root, oblong or lance-shaped. *Flowers.*—White, with two-cleft petals. *Pod.*—Flat, varying from oval to oblong, lance-shaped.

This little plant may be found flowering along the roadsides and in sandy places during April and May. It has come to us from Europe.

SHEPHERD'S PURSE.

Capsella Bursa-pastoris. Mustard Family (p. 17).

Stem.—Low, branching. *Root-leaves.*—Clustered, incised or toothed. *Stem-leaves.* — Arrow-shaped, set close to the stem. *Flowers.* — White, small, in general structure resembling other members of the Mustard family. *Pod.*—Triangular, heart-shaped.

This is one of the commonest of our wayside weeds, working its way everywhere with such persistency and appropriating other people's property so shamelessly, that it has won for itself the nickname of pickpocket. Its popular title arose from the shape of its little seed-pods.

MAY-APPLE. MANDRAKE.

Podophyllum peltatum. Barberry Family.

Flowering-stem.—Two-leaved, one-flowered. *Flowerless-stems.*—Terminated by one large, rounded, much-lobed leaf. *Leaves* (of flowering-stems).—One-sided, five to nine-lobed, the lobes oblong, the leaf-stalks fastened to their lower side near the inner edge. *Flower.*—White, large, nodding from the fork made by the two leaves. *Calyx.*—Of six early-falling sepals. *Corolla.*—Of six to nine rounded petals. *Stamens.*—Twice as many as the petals. *Pistil.*—One, with a large, thick stigma set close to the ovary. *Fruit.*—A large, fleshy, egg-shaped berry, sweet and edible.

" The umbrellas are out ! '' cry the children, when the great green leaves of the May-apple first unfold themselves in spring. These curious-looking leaves at once betray the hiding-place of the pretty but unpleasantly odoriferous flower which nods beneath them. They lie thickly along the woods and meadows in many parts of the country, arresting one's attention by the railways. The fruit, which ripens in July, has been given the name of " wild lemon,'' in some places on account of its shape. It was valued by the Indians for medicinal purposes, and its mawkish flavor still seems to find favor with the children, notwithstanding its frequently unpleasant after-effects. The leaves and roots are poisonous if taken internally, and are said to have been used as a pot herb, with fatal results. They yield an extract which has been utilized in medicine.

TWIN-LEAF. RHEUMATISM-ROOT.

Jeffersonia diphylla. Barberry Family.

A low plant. *Leaves.*—From the root, long-stalked, parted into two rounded leaflets. *Scape.*—One-flowered. *Flower.*—White, one inch broad. *Sepals.*—Four, falling early. *Petals.*—Eight ; flat, oblong. *Stamens.*—Eight. *Pistil.*—One, with a two-lobed stigma.

The twin-leaf is often found growing with the blood-root in the woods of April or May. It abounds somewhat west and southward.

HARBINGER-OF-SPRING.

Erigenia bulbosa. Parsley Family (p. 15).

Stem.—Three to nine inches high, from a deep round tuber. *Leaves.*—One or two, divided into linear-oblong leaf-segments. *Flowers.*—White, small, few, in a leafy-bracted compound umbel.

The pretty little harbinger-of-spring should be easily identified by those who are fortunate enough to find it, for it is one of

PLATE IV

Fruit.

MAY-APPLE —*P. peltatum.*

the smallest members of the Parsley family. It is only common in certain localities, being found in abundance in the neighborhood of Washington, where its flowers appear as early as March.

EARLY EVERLASTING. PLANTAIN-LEAVED EVERLASTING.

Antennaria plantaginifolia. Composite Family (p. 13).

Stems.—Downy or woolly, three to eighteen inches high. *Leaves.*—Silky, woolly when young ; those from the root, oval, three-nerved ; those on the flowering stems, small, lance-shaped. *Flower-heads.*—Crowded, clustered, small, yellowish-white, composed entirely of tubular flowers.

In early spring the hillsides are whitened with this, the earliest of the everlastings.

SPRING BEAUTY.

Claytonia Virginica. Purslane Family.

Stem.—From a small tuber, often somewhat reclining. *Leaves.*—Two ; opposite, long and narrow. *Flowers.*—White, with pink veins, or pink with deeper-colored veins, growing in a loose cluster. *Calyx.*—Of two sepals. *Corolla.*—Of five petals. *Stamens.*—Five. *Pistil.*—One, with style three-cleft at apex.

So bashful when I spied her,
So pretty, so ashamed !
So hidden in her leaflets
Lest anybody find :

So breathless when I passed her,
So helpless when I turned
And bore her struggling, blushing,
Her simple haunts beyond !

For whom I robbed the dingle,
For whom betrayed the dell,
Many will doubtless ask me,
But I shall never tell !

Yet we are all free to guess—and what flower—at least in the early year, before it has gained that touch of confidence which it acquires later—is so bashful, so pretty, so flushed with rosy shame, so eager to defend its modesty by closing its blushing petals when carried off by the despoiler—as the spring beauty ? To be sure, she is not " hidden in her leaflets," although often seeking concealment beneath the leaves of other plants—but why not assume that Miss Dickinson has availed herself of something of the license so freely granted to poets—especially, it seems to me—to poets of nature ? Perhaps of this class few are more accurate than she, and although we wonder at the sudden blindness which leads her to claim that

—Nature rarer uses yellow
Than another hue—

32

PLATE V

SPRING BEAUTY.—*C. Virginica.*

33

when it seems as though it needed but little knowledge of flowers to recognize that yellow, probably, occurs more frequently among them than any other color, and also at the representation of this same nature as

—Spending scarlet like a woman—

when in reality she is so chary of this splendid hue ; still we cannot but appreciate that this poet was in close and peculiar sympathy with flowers, and was wont to paint them with more than customary fidelity.

We look for the spring beauty in April and May, and often find it in the same moist places—on a brook's edge or skirting the wet woods—as the yellow adder's tongue. It is sometimes mistaken for an anemone, but its rose-veined corolla and linear leaves easily identify it. Parts of the carriage-drive in the Central Park are bordered with great patches of the dainty blossoms. One is always glad to discover these children of the country within our city limits, where they can be known and loved by those other children who are so unfortunate as to be denied the knowledge of them in their usual haunts. If the day chances to be cloudy these flowers close and are only induced to open again by an abundance of sunlight. This habit of closing in the shade is common to many flowers, and should be remembered by those who bring home their treasures from the woods and fields, only to discard the majority as hopelessly wilted. If any such exhausted blossoms are placed in the sunlight, with their stems in fresh water, they will probably regain their vigor. Should this treatment fail, an application of very hot—almost boiling—water should be tried. This heroic measure often meets with success.

DUTCHMAN'S BREECHES. WHITE-HEARTS.

Dicentra Cucullaria. Fumitory Family.

Scape.—Slender. *Leaves.*— Thrice - compound. *Flowers.*— White and yellow, growing in a raceme. *Calyx.*—Of two small, scale-like sepals. *Corolla.*—Closed and flattened ; of four somewhat cohering white petals tipped with yellow ; the two outer—large, with spreading tips and deep spurs ; the two inner—small, with spoon-shaped tips uniting over the anthers and stigma. *Stamens.*—Six. *Pistil.*—One.

There is something singularly fragile and spring-like in the appearance of this plant as its heart-shaped blossoms nod from

34

PLATE VI

Tuberous rootstocks.

DUTCHMAN'S BREECHES.—*D. Cucullaria.*

the rocky ledges where they thrive best. One would suppose that the firmly closed petals guarded against any intrusion on the part of insect-visitors and indicated the flower's capacity for self-fertilization ; but it is found that when insects are excluded by means of gauze no seeds are set, which goes to prove that the pollen from another flower is a necessary factor in the continuance of this species. The generic name, *Dicentra*, is from the Greek and signifies *two-spurred*. The flower, when seen, explains its two English titles. It is accessible to every New Yorker, for in early April it whitens many of the shaded ledges in the upper part of the Central Park.

SQUIRREL CORN.

Dicentra Canadensis. Fumitory Family.

The squirrel corn closely resembles the dutchman's breeches. Its greenish or pinkish flowers are heart-shaped, with short, rounded spurs. They have the fragrance of hyacinths, and are found blossoming in early spring in the rich woods of the North.

FOAM-FLOWER. FALSE MITRE-WORT.

Tiarella cordifolia. Saxifrage Family.

Stem.—Five to twelve inches high, leafless, or rarely with one or two leaves. *Leaves.*—From the rootstock or runners, heart-shaped, sharply lobed. *Flowers.*—White, in a full raceme. *Calyx.*—Bell-shaped, five-parted. *Corolla.*—Of five petals on claws. *Stamens.*—Ten, long and slender. *Pistil.*—One, with two styles.

Over the hills and in the rocky woods of April and May the graceful white racemes of the foam flower arrest our attention. This is a near relative of the *Mitella* or true mitre-wort. Its generic name is a diminutive from the Greek for *turban*, and is said to refer to the shape of the pistil.

EARLY SAXIFRAGE.

Saxifraga Virginiensis. Saxifrage Family.

Scape.—Four to nine inches high. *Leaves.*—Clustered at the root, somewhat wedge-shaped, narrowed into a broad leaf-stalk. *Flowers.*—White, small, clustered. *Calyx.*—Five-cleft. *Corolla.*—Of five petals. *Stamens.*—Ten. *Pistil.*—One, with two styles.

In April we notice that the seams in the rocky cliffs and hillsides begin to whiten with the blossoms of the early saxifrage.

PLATE VII

Fruit.

FOAM-FLOWER.—*T. cordifolia.*

37

Steinbrech—stonebreak—the Germans appropriately entitle this little plant, which bursts into bloom from the minute clefts in the rocks and which has been supposed to cause their disintegration by its growth. The generic and common names are from *saxum* —a rock, and *frango*—to break.

MITRE-WORT. BISHOP'S CAP.

Mitella diphylla. Saxifrage Family.

Stem.—Six to twelve inches high, hairy, bearing two opposite leaves. *Leaves.*—Heart-shaped, lobed and toothed, those of the stem opposite and nearly sessile. *Flowers.*—White, small, in a slender raceme. *Calyx.*— Short, five-cleft. *Corolla.*—Of five slender petals which are deeply incised. *Stamens.*—Ten, short. *Pistil.*—One, with two styles.

The mitre-wort resembles the foam flower in foliage, but bears its delicate crystal-like flowers in a more slender raceme. It also is found in the rich woods, blossoming somewhat later.

INDIAN POKE. FALSE HELLEBORE.

Veratrum viride. Lily Family.

Root.—Poisonous, coarse and fibrous. *Stem.*—Stout, two to seven feet high, very leafy to the top. *Leaves.*—Broadly oval, pointed, clasping. *Flowers.* — Dull greenish, inconspicuous, clustered. *Perianth.* — Of six spreading sepals. *Stamens.*—Six. *Pistil.*—One, with three styles.

When we go to the swampy woods in March or April we notice an array of green, solid-looking spears which have just appeared above the ground. If we handle one of these we are impressed with its firmness and rigidity. When the increasing warmth and sunshine have tempted the veiny, many-plaited leaves of the false hellebore to unfold themselves it is difficult to realize that they composed that sturdy tool which so effectively tunnelled its way upward to the earth's surface. The tall stems and large bright leaves of this plant are very noticeable in the early year, forming conspicuous masses of foliage while the trees and shrubs are still almost leafless. The dingy flowers which appear later rarely attract attention.

CARRION-FLOWER. CAT-BRIER.

Smilax herbacea. Lily Family.

Stem.—Climbing, three to fifteen feet high. *Leaves.*—Ovate, or rounded heart-shaped, or abruptly cut off at base, shining. *·Flowers.*—Greenish or yellowish, small, clustered, unisexual. *Perianth.*—Six-parted. *Stamens.*—six. *Pistil.*—One, with three spreading stigmas. (Stamens and pistils occurring on different plants.) *Fruit.*—A bluish-black berry.

One whiff of the foul breath of the carrion flower suffices for its identification. Thoreau likens its odor to that of "a dead rat in the wall." It seems unfortunate that this strikingly handsome plant which clambers so ornamentally over the luxuriant thickets which border our lanes and streams, should be so handicapped each June. Happily with the disappearance of the blossoms, it takes its place as one of the most attractive of our climbers.

The common green-brier, *S. rotundifolia*, is a near relation which is easily distinguished by its prickly stem.

The dark berries and deeply tinted leaves of this genus add greatly to the glorious autumnal display along our roadsides and in the woods and meadows.

LARGER WHITE TRILLIUM.

Trillium grandiflorum. Lily Family.

Stem.—Stout, from a tuber-like rootstock. *Leaves.*—Ovate, three in a whorl, a short distance below the flower. *Flower.*—Single, terminal, large, white, turning pink or marked with green. *Calyx.*—Of three green, spreading sepals. *Corolla.*—Of three long pointed petals. *Stamens.*—Six.—*Pistil.*—One, with three spreading stigmas. *Fruit.*—A large ovate, somewhat angled, red berry.

This very beautiful and decorative flower must be sought far from the highway in the cool rich woods of April and May. Mr. Ellwanger speaks of the "chaste pure triangles of the white wood-lily," and says that it often attains a height of nearly two feet.

T. cernuum has no English title. Its smaller white or pinkish blossom is borne on a stalk which is so much curved as to sometimes quite conceal the flower beneath the leaves. It may be sought in the moist places in the woods.

The painted trillium, *T. erythrocarpum*, is also less large and showy than the great white trillium, but it is quite as pleasing. Its white petals are painted at their base with red stripes. This species is very plentiful in the Adirondack and Catskill Mountains.

GROUND-NUT. DWARF GINSENG.

Aralia trifolia. Ginseng Family.

Stem.—Four to eight inches high. *Leaves.*—Three in a whorl, divided into from three to five leaflets. *Flowers.*—White, in an umbel. *Fruit.*—Yellowish, berry-like. *Root.*—A globular tuber.

The tiny white flowers of the dwarf ginseng are so closely clustered as to make " one feathery ball of bloom," to quote Mr. Hamilton Gibson. This little plant resembles its larger relative, the true ginseng. It blossoms in our rich open woods early in spring, and hides its small round tuber so deep in the earth that it requires no little care to uproot it without breaking the slender stem. This tuber is edible and pungent-tasting, giving the plant its name of ground-nut.

GINSENG.

Aralia quinquefolia. Ginseng Family.

Root.—Large and spindle-shaped, often forked. *Stem.*—About one foot high. *Leaves.*—Three in a whorl, divided into leaflets. *Flowers.*—Greenish-white, in a simple umbel. *Fruit.*—Bright red, berry-like.

This plant is well known by name, but is yearly becoming more scarce. The aromatic root is so greatly valued in China for its supposed power of combating fatigue and old age that it can only be gathered by order of the emperor. The forked specimens are believed to be the most powerful, and their fancied likeness to the human form has obtained for the plant the Chinese title of *Jin-chen* (from which ginseng is a corruption), and the Indian one of *Garan-toguen*, both of which, strangely enough, are said to signify, *like a man.* The Canadian Jesuits first began to ship the roots of the American species to China, where they sold at about five dollars a pound. At present they are said to command about one-fifth of that price in the home market.

PLATE VIII

Fruit.

PAINTED TRILLIUM.—*T. erythrocarpum.*

WILD SARSAPARILLA.

Aralia nudicaulis. Ginseng Family.

Stem.—Bearing a single large, long-stalked, much-divided leaf, and a shorter naked scape which bears the rounded flower-clusters. *Flowers.*— Greenish-white, in umbels. *Calyx.*—With short or obsolete teeth. *Corolla.* —Of five petals. *Stamens.*—Five. *Fruit.*—Black or dark-purple, berry-like.

In the June woods the much-divided leaf and rounded flower-clusters of the wild sarsaparilla are frequently noticed, as well as the dark berries of the later year. The long aromatic roots of this plant are sold as a substitute for the genuine sarsaparilla. The rice-paper plant of China is a member of this genus.

SPIKENARD.

Aralia racemosa. Ginseng Family.

Root.—Large and aromatic. *Stem.*—Often tall and widely branched, leafy. *Leaves.*—Divided into many leaflets. *Flowers.* Greenish-white, in clusters which are racemed. *Fruit.*—Dark purple, berry-like.

CANADA VIOLET.

Viola Canadensis. Violet Family.

Stem.—Leafy, upright, one to two feet high. *Leaves.*—Heart-shaped, pointed, toothed. *Flowers.*—White, veined with purple, violet beneath, otherwise greatly resembling the common blue violet.

We associate the violet with the early year, but I have found the delicate fragrant flowers of this species blossoming high up on the Catskill Mountains late into September ; and have known them to continue to appear in a New York city-garden into November. They are among the loveliest of the family, having a certain sprightly self-assertion which is peculiarly charming, perhaps because so unexpected.

The tiny sweet white violet, *V. blanda*, with brown or purple veins, which is found in nearly all low, wet, woody places in spring, is perhaps the only uniformly fragrant member of the family, and its scent, though sweet, is faint and elusive.

The lance-leaved violet, *V. lanceolata*, is another white

PLATE IX

Flower.   Fruit.

WILD SARSAPARILLA.—*A. nudicaulis.*

43

species which is easily distinguished by its smooth lance-shaped leaves, quite unlike those of the common violet. It is found in damp soil, especially eastward.

SOLOMON'S SEAL.

Polygonatum biflorum. Lily Family.

Stem.—Slender, curving, one to three feet long. *Leaves.*—Alternate, oval, set close to the stem. *Flowers.*—Greenish-white or straw-colored, bell-shaped, nodding from the axils of the leaves. *Perianth.*—Six-lobed at the summit. *Stamens.*—Six. *Pistil.*—One. *Fruit.* A dark blue berry.

The graceful leafy stems of the Solomon's seal are among the most decorative features of our spring woods. The small blossoms which appear in May grow either singly or in clusters on a flower-stalk which is so fastened into the axil of each leaf that they droop beneath, forming a curve of singular grace which is sustained in later summer by the dark blue berries.

The larger species, *P. giganteum*, grows to a height of from two to seven feet, blossoming in the meadows and along the streams in June.

The common name was suggested by the rootstocks, which are marked with large round scars left by the death and separation of the base of the stout stalks of the previous years. These scars somewhat resemble the impression of a seal upon wax.

The generic name is from two Greek words signifying *many*, and *knee*, alluding to the numerous joints of the rootstock.

CHOKE-BERRY.

Pyrus arbutifolia. Rose Family.

A shrub from one to three feet high. *Leaves.*—Oblong or somewhat lance-shaped, finely toothed, downy beneath. *Flowers.*—White or reddish, small, clustered. *Calyx.*—Five-cleft. *Corolla.*—Of five petals. *Stamens.* —Numerous. *Pistil.*—One, with two to five styles. *Fruit.*—Small, pear-shaped or globular, berry-like, dark red or blackish.

This low shrub is common in swamps and moist thickets all along the Atlantic coast, as well as farther inland. Its flowers appear in May or June ; its fruit in late summer or autumn.

PLATE X

Rootstock.

SOLOMON'S SEAL.—*P. biflorum.*

CREEPING SNOWBERRY.

Chiogenes serpyllifolia. Heath Family.

Stem.—Slender, trailing and creeping. *Leaves.* — Evergreen, small, ovate, pointed. *Flowers.*—Small, white, solitary from the axils of the leaves. *Calyx.*—Four-parted, with four large bracelets beneath. *Corolla.* —Deeply four-parted. *Stamens.*—Eight. *Pistil.*—One. *Fruit.*—A pure white berry.

This pretty little creeper is found blossoming in May in the peat-bogs and mossy woods of the North. It is only conspicuous when hung with its snow-white berries in late summer. It has the aromatic flavor of the wintergreen.

BEARBERRY.

Arctostaphylos Uva-ursi. Heath Family.

A trailing shrub. *Leaves.*—Thick and evergreen, smooth, somewhat wedge-shaped. *Flowers.*—Whitish, clustered. *Calyx.*—Small. *Corolla.* — Urn-shaped, five-toothed. *Stamens.* — Ten. *Pistil.* — One. *Fruit.*— Red, berry-like.

This plant blossoms in May, and is found on rocky hillsides. Its name refers to the relish with which bears are supposed to devour its fruit.

FALSE SOLOMON'S SEAL.

Smilacina racemosa. Lily Family.

Stem.—Usually curving, one to three feet long. *Leaves.*—Oblong, veiny. *Flowers.*—Greenish-white, small, in a terminal raceme.—*Perianth.*—Six-parted. *Stamens.*—Six. *Pistil.*—One. *Fruit.*—A pale red berry speckled with purple.

A singular lack of imagination is betrayed in the common name of this plant. Despite a general resemblance to the true Solomon's seal, and the close proximity in which the two are constantly found, *S. racemosa*, has enough originality to deserve an individual title. The position of the much smaller flowers is markedly different. Instead of drooping beneath the stem they terminate it, having frequently a pleasant fragrance, while the berries of late summer are pale red, flecked with purple. It puzzles one to understand why these two plants should so constantly be found growing side by side—so close at times that they almost appear to spring from one point. The generic name is

PLATE XI

Single flower.   Fruit.

FALSE SOLOMON'S SEAL.—*S. racemosa.*

47

from *smilax*, on account of a supposed resemblance between the leaves of this plant and those which belong to that genus.

MAPLE-LEAVED VIBURNUM. DOCKMACKIE. ARROW-WOOD.

Viburnum acerifolium. Honeysuckle Family.

A shrub from three to six feet high. *Leaves.*—Somewhat three-lobed, resembling those of the maple, downy underneath. *Flowers.*—White, small, in flat-topped clusters. *Calyx.*—Five-toothed. *Corolla.*—Spreading, five-lobed. *Stamens.*—Five. *Pistil.*—One. *Fruit.*—Berry-like, crimson, turning purple.

Perhaps our flowering shrubs contribute even more to the beauty of the early-summer woods and fields than the smaller plants. Along many of the lanes which intersect the woodlands the viburnums are conspicuous in June. When the blossoms of the dockmackie have passed away we need not be surprised if we are informed that this shrub is a young maple. There is certainly a resemblance between its leaves and those of the maple, as the specific name indicates. To be sure, the first red, then purple berries, can scarcely be accounted for, but such a trifling incongruity would fail to daunt the would-be wiseacre of field and forest. With Napoleonic audacity he will give you the name of almost any shrub or flower about which you may inquire. Seizing upon some feature he has observed in another plant, he will immediately christen the one in question with the same title— somewhat modified, perhaps—and in all probability his authority will remain unquestioned. There is a marvellous amount of inaccuracy afloat in regard to the names of even the commonest plants, owing to this wide-spread habit of guessing at the truth and stating a conjecture as a fact.

HOBBLE-BUSH. AMERICAN WAYFARING-TREE.

Viburnum lantanoides. Honeysuckle Family.

Leaves.—Rounded, pointed, closely toothed, heart-shaped at the base, the veins beneath as well as the stalks and small branches being covered with a rusty scurf. *Fruit.*—Coral-red, berry-like.

The marginal flowers of the flat-topped clusters of the hobble-bush, like those of the hydrangea, are much larger than the inner ones, and are without either stamens or pistils; their only part in the economy of the shrub being to form an attractive setting for the cluster, and thus to allure the insect-visitors that are usually so necessary to the future well-being of the species. The shrub is a common one in our northern woods and mountains. Its straggling growth, and its reclining branches, which often take root in the ground, have suggested the popular names of hobble-bush, and wayfaring-tree.

ROUND-LEAVED DOGWOOD.

Cornus circinata. Dogwood Family.

A shrub six to ten feet high. *Leaves.*—Rounded, abruptly pointed. *Flowers.*—Small, white, in flat, spreading clusters. *Calyx.*—Minutely four-toothed. *Corolla.*—Of four white, oblong, spreading petals. *Stamens.*—Four. *Pistil.*—One. *Fruit.*—Light blue, berry-like.

The different members of the Dogwood family are important factors in the lovely pageant which delights our eyes along the country lanes every spring. Oddly enough, only the smallest and largest representative of the tribe (the little bunch-berry, and the flowering-dogwood, which is sometimes a tree of goodly dimensions), have in common the showy involucre which is usually taken for the blossom itself; but which instead only surrounds the close cluster of inconspicuous greenish flowers.

·The other members of the genus are all comprised in the shrubby dogwoods; many of these are very similar in appearance, bearing their white flowers in flat, spreading clusters, and differing chiefly in their leaves and fruit.

The branches of the round-leaved dogwood are greenish and warty-dotted. Its fruit is light blue, and berry-like.

The panicled dogwood, *C. paniculata,* may be distinguished by its white fruit and smooth, gray branches.

The red-osier dogwood, *C. stolonifera,* is common in wet places. Its young shoots and branches are a bright purplish-red. Its flower-clusters are small; its fruit, white or lead-color.

The bark of this genus has been considered a powerful tonic, and an extract entitled " cornine," is said to possess the properties of quinine less strongly marked. The Chinese peel its twigs, and use them for whitening their teeth. It is said that the Creoles also owe the dazzling beauty of their teeth to this same practice.

BELLWORT.

Oakesia sessilifolia. Lily Family.

Stem.—Acutely angled, rather low. *Leaves.*—Set close to or clasping the stem, pale, lance-oblong. *Flower.*—Yellowish-white or straw-color. *Perianth.*—Narrowly bell-shaped, divided into six distinct sepals. *Stamens.* —Six. *Pistil.*—One, with a deeply three-cleft style.

In spring this little plant is very abundant in the woods. It bears one or two small lily-like blossoms which droop modestly beneath the curving stems.

With the same common name and near of kin is *Uvularia perfoliata*, with leaves which seem pierced by the stem, but otherwise of a strikingly similar aspect.

HAWTHORN. WHITE-THORN.

Cratægus coccinea. Rose Family.

A shrub or small tree, with spreading branches, and stout thorns or spines. *Leaves.*—On slender leaf-stalks, thin, rounded, toothed, sometimes lobed. *Flowers.*—White or sometimes reddish, rather large, clustered, with a somewhat disagreeable odor. *Calyx.*—Urn-shaped, five-cleft. *Corolla.*— Of five broad, rounded petals. *Stamens.*—Five to ten, or many. *Pistil.*— One, with one to five styles. *Fruit.*—Coral-red, berry-like.

The flowers of the white-thorn appear in spring, at the same time with those of the dogwoods. Its scarlet fruit gleams from the thicket in September.

WHITE BANEBERRY.

Actæa alba. Crowfoot Family.

Stem.—About two feet high. *Leaves.*—Twice or thrice-compound, leaf-lets incised and sharply toothed. *Flowers.*—Small, white, in a thick, oblong, terminal raceme. *Calyx.*—Of four to five tiny sepals which fall as the flower expands. *Corolla.*—Of four to ten small flat petals with slender claws. *Stamens.*—Numerous, with slender white filaments. *Pistil.*—One, with a depressed, two-lobed stigma. *Fruit.*—An oval white berry, with a dark spot, on a *thick red stalk.*

The feathery clusters of the white baneberry may be gathered when we go to the woods for the columbine, the wild ginger,

PLATE XII

Fruit.   Fruit.

Oakesia sessilifolia.   *U. perfoliata.*

BELLWORT.

51

the Jack-in-the-pulpit, and Solomon's seal. These flowers are very nearly contemporaneous and seek the same cool shaded nooks, all often being found within a few feet of one another.

The red baneberry, *A. rubra*, is a somewhat more Northern plant and usually blossoms a week or two earlier. Its cherry-red (occasionally white) berries on their *slender stalks* are easily distinguished from the white ones of *A. alba*, which look strikingly like the china eyes that small children occasionally manage to gouge from their dolls' heads.

MOUNTAIN HOLLY.

Nemopanthes fascicularis. Holly Family.

A much-branched shrub, with ash-gray bark. *Leaves.*—Alternate, oblong, smooth, on slender leaf-stalks. *Flowers.*—White, some perfect, others unisexual ; solitary or clustered in the axils of the leaves on long, slender flower-stalks. *Calyx.*—Minute or obsolete. *Corolla.*—Of four or five spreading petals. *Stamens.*—Four or five. *Pistil.*—One. *Fruit.*—Coral-red, berry-like.

The flowers of this shrub appear in the damp woods of May. Its light red berries on their slender stalks are noticed in late summer when its near relation, the black alder or winterberry is also conspicuous. Its generic name signifies *flower with a thread-like stalk.*

WINTERBERRY. BLACK ALDER.

Ilex verticillata. Holly Family.

A shrub, common in low grounds. *Leaves.*—Oval or lance-shaped, pointed at apex and base, toothed. *Flowers.*—White ; some perfect, others unisexual ; clustered on very short flower-stalks in the axils of the leaves ; appearing in May or June. *Calyx.*—Minute. *Corolla.*—Of four to six petals. *Stamens.*—Four to six. *Pistil.*—One. *Fruit.*—Coral-red, berry-like.

The year may draw nearly to its close without our attention being arrested by this shrub. But in September it is well nigh impossible to stroll through the country lanes without pausing to admire the bright red berries clustered so thickly among the leaves of the black alder. The American holly, *I. opaca*, is

PLATE XIII

Fruit.

WHITE BANEBERRY.—*A. alba.*

53

closely related to this shrub, whose generic name is the ancient Latin title for the holly-oak.

RED-BERRIED ELDER.

Sambucus racemosa. Honeysuckle Family.

Stems.—Woody, two to twelve feet high. *Leaves.*—Divided into leaf- lets. *Flowers.*—White, resembling those of the Common Elder (p. 78), but borne in pyramidal instead of in flat-topped clusters. *Fruit.*—Bright red, berry-like.

The white clusters of the red-berried elder are found in the rocky woods of May ; its scarlet fruit, like that of the shad- bush, appearing in June.

BUNCH-BERRY. DWARF CORNEL.

Cornus Canadensis. Dogwood Family.

Stem.—Five to seven inches high.—*Leaves.*—Ovate, pointed, the upper crowded into an apparent whorl of four to six. *Flowers.*—Greenish, small, in a cluster which is surrounded by a large and showy four-leaved, petal-like, white or pinkish involucre. *Calyx.*—Minutely four-toothed. *Corolla.*—Of four spreading petals. *Stamens.*—Four. *Pistil.*—One. *Fruit.*—Bright red, berry-like.

When one's eye first falls upon the pretty flowers of the bunch-berry in the June woods, the impression is received that each low stem bears upon its summit a single large white blossom. A more searching look discovers that what appeared like rounded petals are really the showy white leaves of the involucre which surround the small, closely clustered, greenish flowers.

The bright red berries which appear in late summer make brilliant patches in the woods and swamps. Both in flower and fruit this is one of the prettiest of our smaller plants. It is closely allied to the well-known flowering-dogwood, which is so ornamental a tree in early spring.

In the Scotch Highlands it is called the "plant of glut- tony," on account of its supposed power of increasing the appe- tite. It is said to form part of the winter food of the Esqui- maux.

PLATE XIV

Fruit.

BUNCH-BERRY.—*C. Canadensis.*

55

SWEET BAY. LAUREL MAGNOLIA.

Magnolia glauca. Magnolia Family.

A shrub from four to twenty feet high. *Leaves.*—Oval to broadly lance-shaped, from three to six inches long. *Flowers.*—White, two inches long, growing singly at the ends of the branches. *Calyx.*—Of three sepals. *Corolla.*—Globular, with from six to nine broad petals. *Stamens.*—Numerous, with short filaments and long anthers. *Pistils.*—Many, packed so as to make a sort of cone in fruit. *Fruit.*—Cone-like, red, fleshy when ripe ; the pistils opening at maturity and releasing the scarlet seeds which hang by delicate threads.

The beautiful fragrant blossoms of the sweet bay may be found from June till August, in swamps along the coast from Cape Ann southward.

LIZARD'S TAIL.

Saururus cernuus. Pepper Family.

Stem.—Jointed, often tall. *Leaves.*—Alternate, heart-shaped. *Flowers.*—White, without calyx or corolla, crowded into a slender, wand-like terminal spike which nods at the end. *Stamens.*—Usually six or seven. *Pistils.*—Three or four, united at their base.

The nodding, fragrant spikes of the lizard's tail abound in certain swamps from June till August. While the plant is not a common one, it is occasionally found in great profusion, and is sure to arrest attention by its odd appearance.

MOONSEED.

Menispermum Canadense. Moonseed Family.

Stem.—Woody, climbing. *Leaves.*—Three to seven-angled or lobed, their stalks fastened near the edge of the lower surface. *Flowers.*—White or yellowish, in small loose clusters, unisexual. *Calyx.*—Of four to eight sepals. *Corolla.*—Of six to eight short petals. *Stamens and Pistils.*—Occurring on different plants. *Fruit.*—Berry-like, black, with a bloom.

Clambering over the thickets which line the streams, we notice in September the lobed or angled leaves and black berries of the moonseed, the small white or yellowish flowers of which were, perhaps, overlooked in June.

MOUNTAIN LAUREL. SPOONWOOD. CALICO-BUSH.

Kalmia latifolia. Heath Family.

An evergreen shrub. *Leaves.*—Oblong, pointed, shining, of a leathery texture. *Flowers.*—White or pink, in terminal clusters. *Calyx.*—Five-parted. *Corolla.*—Marked with red, wheel-shaped, five-lobed, with ten depressions. *Stamens.*—Ten, each anther lodged in one of the depressions of the corolla. *Pistil.*—One.

The shining green leaves which surround the white or rose-colored flowers of the mountain laurel are familiar to all who have skirted the west shore of the Hudson River, wandered across the hills that lie in its vicinity, or clambered across the mountains of Pennsylvania, where the shrub sometimes grows to a height of thirty feet. Not that these localities limit its range : for it abounds more or less from Canada to Florida, and far inland, especially along the mountains, whose sides are often clothed with an apparent mantle of pink snow during the month of June, and whose waste places are, in very truth, made to blossom like the rose at this season.

The shrub is highly prized and carefully cultivated in England. Barewood Gardens, the beautiful home of the editor of the London *Times*, is celebrated for its fine specimens of mountain laurel and American rhododendron. The English papers advertise the approach of the flowering season, the estate is thrown open to the public, and the people for miles around flock to see the radiant strangers from across the water. The shrub is not known there as the laurel, but by its generic title, *Kalmia.* ' The head gardener of the place received with some incredulity , my statement that in parts of America the waste hill-sides were brilliant with its beauty every June.

The ingenious contrivance of these flowers to secure cross-fertilization is most interesting. The long filaments of the stamens are arched by each anther being caught in a little pouch of the corolla ; the disturbance caused by the sudden alighting of an insect on the blossom, or the quick brush of a bee's wing, dislodges the anthers from their niches, and the stamens spring upward with such violence that the pollen is jerked from its hiding-place in the pore of the anther-cell on to the body of the insect-visitor, who straightway carries it off to another flower upon whose protruding stigma it is sure to be inadvertently deposited. In order to see the working of this for one's self, it is only necessary to pick a fresh blossom and either brush the corolla quickly with one's finger, or touch the stamens suddenly with a pin, when the anthers will be dislodged and the pollen will be seen to fly.

This is not the laurel of the ancients—the symbol of victory and fame—notwithstanding some resemblance in the form of the leaves. The classic shrub is supposed to be identical with the *Laurus nobilis* which was carried to our country by the early colonists, but which did not thrive in its new environment.

The leaves of our species are supposed to possess poisonous qualities, and are said to have been used by the Indians for suicidal purposes. There is also a popular belief that the flesh of a partridge which has fed upon its fruit becomes poisonous. The clammy exudation about the flower-stalks and blossoms may serve the purpose of excluding from the flower such small insects as would otherwise crawl up to it, dislodge the stamens, scatter the pollen, and yet be unable to carry it to its proper destination on the pistil of another flower.

The *Kalmia* was named by Linnæus after Peter Kalm, one of his pupils who travelled in this country, who was, perhaps, the first to make known the shrub to his great master.

The popular name spoonwood grew from its use by the Indians for making eating-utensils. The wood is of fine grain and takes a good polish.

The title calico-bush probably arose from the marking of the corolla, which, to an imaginative mind, might suggest the cheap cotton-prints sold in the shops.

WHITE SWAMP HONEYSUCKLE. CLAMMY AZALEA.
Rhododendron viscosum. Heath Family.

A shrub from three to ten feet high. *Leaves.*—Oblong. *Flowers.*—White, clustered, appearing after the leaves. *Calyx-lobes.*—Minute. *Corolla.*—White, five-lobed, the clammy tube much longer than the lobes. *Stamens.*—Usually five, protruding. *Pistil.*—One, protruding.

The fragrant white flowers of this beautiful shrub appear in early summer along the swamps which skirt the coast, and occa-

PLATE XV

MOUNTAIN LAUREL.—*K. latifolia.*

sionally farther inland. The close family resemblance to the pink azalea (Pl. LXV) will be at once detected. On the branches of both species will be found those abnormal, fleshy growths, called variously "swamp apples" and "May apples," which are so relished by the children. Formerly these growths were attributed to the sting of an insect, as in the "oak apple;" now they are generally believed to be modified buds.

AMERICAN RHODODENDRON. GREAT LAUREL.

Rhododendron maximum. Heath Family.

A shrub from six to thirty-five feet high. *Leaves.*— Thick and leathery, oblong, entire. *Flowers.*—White or pink, clustered. *Calyx.*—Minute, five-toothed. *Corolla.*—Somewhat bell-shaped, five-parted, greenish in the throat, with red, yellow, or green spots. *Stamens.*—Usually ten. *Pistil.*—One.

This beautiful native shrub is one of the glories of our country when in the perfection of its loveliness. The woods which nearly cover many of the mountains of our Eastern States hide from all but the bold explorer a radiant display during the early part of July. Then the lovely waxy flower-clusters of the American rhododendron are in their fulness of beauty. As in the laurel, the clammy flower-stalks seem fitted to protect the blossom from the depredations of small and useless insects, while the markings on the corolla attract the attention of the desirable bee.

In those parts of the country where it flourishes most luxuriantly, veritable rhododendron jungles termed "hells" by the mountaineers are formed. The branches reach out and interlace in such a fashion as to be almost impassable.

The nectar secreted by the blossoms is popularly supposed to be poisonous. We read in Xenophon that during the retreat of the Ten Thousand, the soldiers found a quantity of honey of which they freely partook, with results that proved almost fatal. This honey is said to have been made from a rhododendron which is still common in Asia Minor and which is believed to possess intoxicating and poisonous properties.

Comparatively little attention had been paid to this superb flower until the Centennial Celebration at Philadelphia, when

PLATE XVI

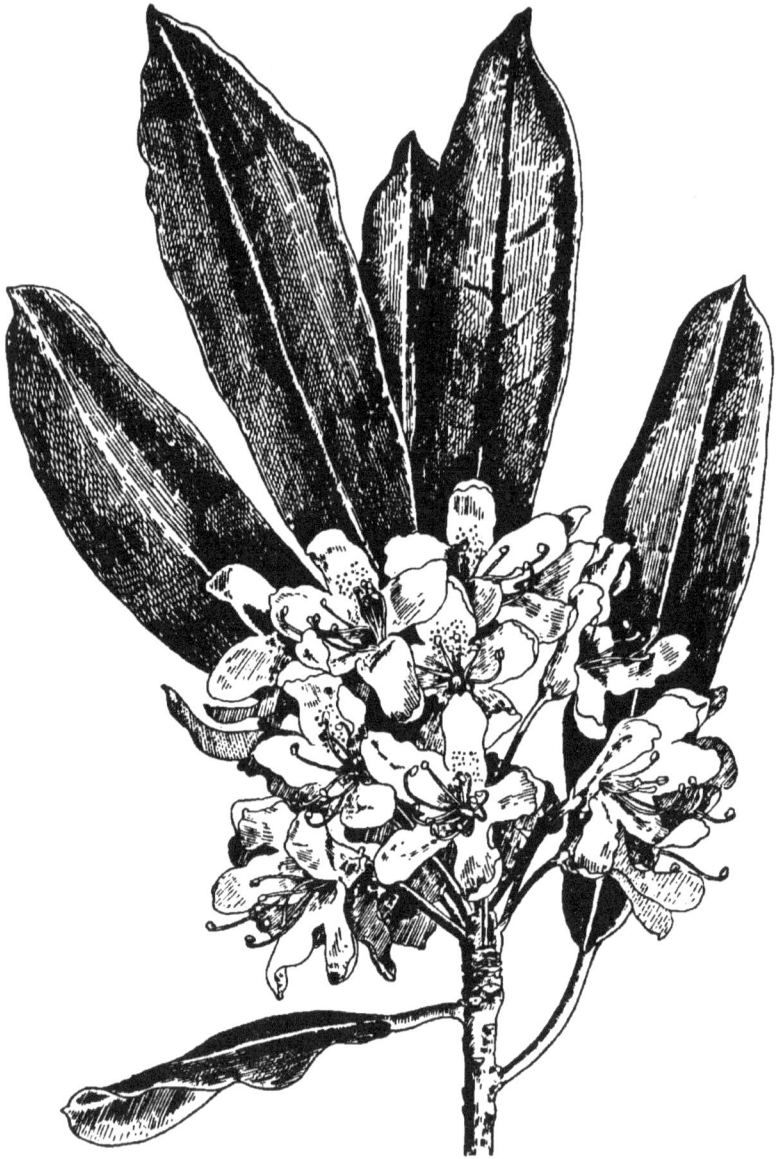

AMERICAN RHODODENDRON.—*R. maximum.*

some fine exhibits attracted the admiration of thousands. The shrub has been carefully cultivated in England, having been brought to great perfection on some of the English estates. It is yearly winning more notice in this country.

The generic name is from the Greek for *rose-tree*.

WOOD SORREL.

Oxalis Acetosella. Geranium Family.

Scape.—One-flowered, two to five inches high. *Leaves.*—Divided into three clover - like leaflets. *Flower.*—White veined with red, solitary. *Calyx.*—Of five sepals. *Corolla.*—Of five petals. *Stamens.*—Ten. *Pistil.* —One with five styles.

Surely nowhere can be found a daintier carpeting than that made by the clover-like foliage of the wood sorrel when studded with its rose-veined blossoms in the northern woods of June. At the very name comes a vision of mossy nooks where the sunlight only comes on sufferance, piercing its difficult path through the tent-like foliage of the forest, resting only long enough to become a golden memory.

The early Italian painters availed themselves of its chaste beauty. Mr. Ruskin says: " Fra Angelico's use of the *Oxalis Acetosella* is as faithful in representation as touching in feeling. The triple leaf of the plant and white flower stained purple probably gave it strange typical interest among the Christian painters."

Throughout Europe it bears the odd name of " Hallelujah " on account of its flowering between Easter and Whitsuntide, the season when the Psalms sung in the churches resound with that word. There has been an unfounded theory that this title sprang from St. Patrick's endeavor to prove to his rude audience the possibility of a Trinity in Unity from the three-divided leaves. By many this ternate leaf has been considered the shamrock of the ancient Irish.

The English title, " cuckoo-bread," refers to the appearance of the blossoms at the season when the cry of the cuckoo is first heard.

Our name sorrel is from the Greek for *sour* and has reference

PLATE XVII

WOOD SORREL.—*O. Acetosella.*

63

to the acrid juice of the plant. The delicate leaflets "sleep" at night; that is, they droop and close one against another.

POISON SUMACH.

Rhus venenata. Cashew Family.

A shrub from six to eighteen feet high. *Leaves.*—Divided into seven to thirteen oblong entire leaflets. *Flowers.*—Greenish or yellowish-white, in loose axillary clusters ; some perfect, others unisexual. *Fruit.*—Whitish or dun-colored, small, globular.

The poison sumach infests swampy places and flowers in June. In early summer it can be distinguished from the harmless members of the family by the slender flower-clusters which grow from the axils of the leaves, those of the innocent sumachs being borne in pyramidal, terminal clusters. In the later year the fruits of the respective shrubs are, of course, similarly situated, but, to accentuate the distinction, they differ in color ; that of the poison sumach being whitish or dun-colored, while that of the other is crimson.

STAGHORN SUMACH.

Rhus typhina. Cashew Family.

A shrub or tree from ten to thirty feet high. *Leaves.*—Divided into eleven to thirty-one somewhat lance-shaped, toothed leaflets. *Flowers.*— Greenish or yellowish-white, in upright terminal clusters, some perfect, others unisexual, appearing in June. *Fruit.*—Crimson, small, globular, hairy.

This is the common sumach which illuminates our hill-sides every autumn with masses of flame-like color. Many of us would like to decorate our homes with its brilliant sprays, but are deterred from handling them by the fear of being poisoned, not knowing that one glance at the crimson fruit-plumes should reassure us, as the poisonous sumachs are white-fruited. These tossing pyramidal fruit-clusters at first appear to explain the common title of staghorn sumach. It is not till the foliage has disappeared, and the forked branches are displayed in all their nakedness, that we feel that these must be the feature in which the common name originated.

POISON IVY.

Rhus Toxicodendron. Cashew Family.

A shrub which usually climbs by means of rootlets over rocks, walls, and trees; sometimes low and erect. *Leaves.*—Divided into three somewhat four-sided pointed leaflets. *Flowers.*—Greenish or yellowish-white, small, some perfect, others unisexual; in loose clusters in the axils of the leaves in June. *Fruit.*—Small, globular, somewhat berry-like, dun-colored, clustered.

This much-dreaded plant is often confused with the beautiful Virginia creeper, occasionally to the ruthless destruction of the latter. Generally the two can be distinguished by the three-divided leaves of the poison ivy, the leaves of the Virginia creeper usually being five-divided. In the late year the whitish fruit of the ivy easily identifies it, the berries of the creeper being blackish. The poison ivy is reputed to be especially harmful during the night, or at any time in early summer when the sun is not shining upon it.

VIRGINIA CREEPER. AMERICAN IVY.

Ampelopsis quinquefolia. Vine Family.

A woody vine climbing by means of disk-bearing tendrils, and also by rootlets. *Leaves.*—Usually divided into five leaflets. *Flowers.*—Greenish, small, clustered, appearing in July. *Fruit.*—A small, blackish berry in October.

Surely in autumn, if not always, this is the most beautiful of our native climbers. At that season its blood-like sprays are outlined against the dark evergreens about which they delight to twine, showing that marvellous discrimination in background which so constantly excites our admiration in nature. The Virginia creeper is extensively cultivated in Europe. Even in Venice, that sea-city where one so little anticipates any reminders of home woods and meadows, many a dim canal mirrors in October some crumbling wall or graceful trellis aglow with its vivid beauty.

SHIN-LEAF.

Pyrola elliptica. Heath Family.

Scape.—Upright, scaly, terminating in a many-flowered raceme. *Leaves.*—From the root, thin and dull, somewhat oval. *Flowers.*—White, nodding. *Calyx.*—Five-parted. *Corolla.*—Of five rounded, concave petals. *Stamens.*—Ten. *Pistil.*—One, with a long curved style.

In the distance these pretty flowers suggest the lilies-of-the-valley. They are found in the woods of June and July, often in close company with the pipsissewa. The ugly common name of shin-leaf arose from an early custom of applying the leaves of this genus to bruises or sores ; the English peasantry being in the habit of calling any kind of plaster a "shin-plaster" without regard to the part of the body to which it might be applied. The old herbalist, Salmon, says that the name *Pyrola* was given to the genus by the Romans on account of the fancied resemblance of its leaves and flowers to those of a pear-tree. The English also call the plant "wintergreen," which name we usually reserve for *Gaultheria procumbens.*

P. rotundifolia is a species with thick, shining, rounded leaves.

COMMON BLACK HUCKLEBERRY.

Gaylussacia resinosa. Heath Family.

One to three feet high. *Stems.*—Shrubby, branching. *Leaves.*—Oval or oblong, sprinkled more or less with waxy, resinous atoms. *Flowers.*—White, reddish, or purplish, bell-shaped, growing in short, one-sided clusters. *Calyx.*—With five short teeth. *Corolla.*—Bell-shaped, with a five-cleft border. *Stamens.*—Ten. *Pistil.*—One. *Fruit.*—A black, bloomless, edible berry.

The flowers of the common huckleberry appear in May or June ; the berries in late summer. The shrub abounds in rocky woods and swamps.

COMMON BLUEBERRY.

Vaccinium corymbosun. Heath Family.

Five to ten feet high.

The blueberry has a bloom which is lacking in the huckleberry. It is found in swamps or low thickets in late summer.

PLATE XVIII

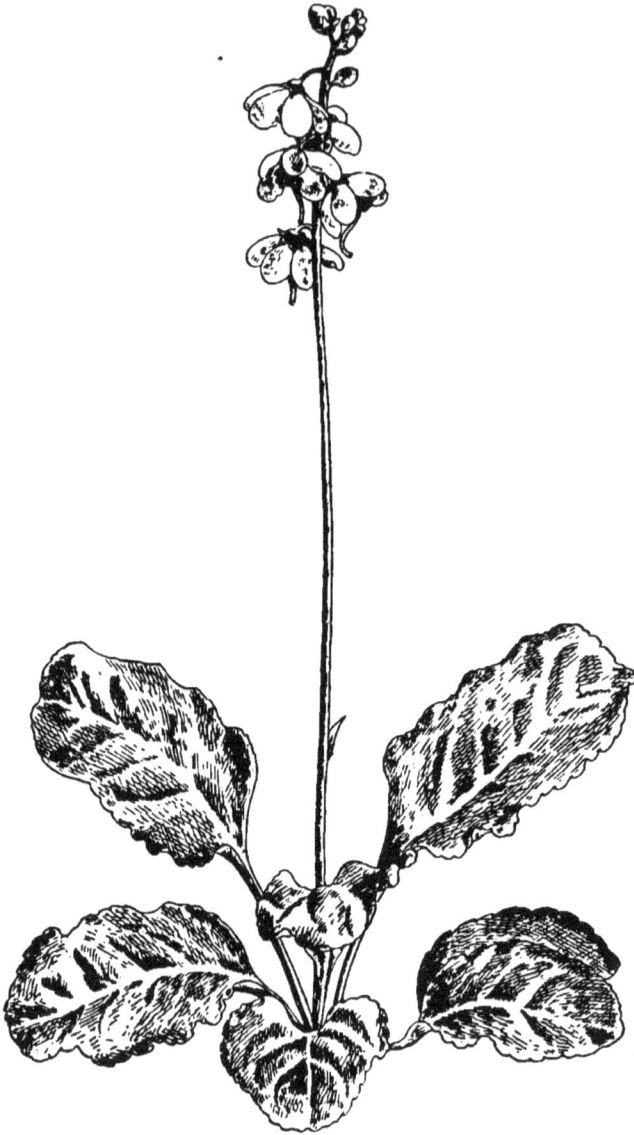

SHIN-LEAF.—*P. elliptica.*

SQUAW HUCKLEBERRY.

Vaccinium stamineum. Heath Family.

Two or three feet high. *Stems.*—Diffusely branched.

This large greenish or yellowish berry is hardly edible. The flowers appear in June, and are easily recognized by their protruding stamens. The leaves are pale green above and whitish underneath.

PIPSISSEWA. PRINCE'S PINE.

Chimaphila umbellata. Heath Family.

Stem.—Four to ten inches high, leafy. *Leaves.*—Somewhat whorled or scattered, evergreen, lance-shaped, with sharply toothed edges. *Flowers.*—White or purplish, fragrant, in a loose terminal cluster. *Calyx.*—Five-lobed. *Corolla.*—With five rounded, widely spreading petals. *Stamens.*—Ten, with violet anthers. *Pistil.*—One, with a short top-shaped style and disk-like stigma.

When strolling through the woods in summer one is apt to chance upon great patches of these deliciously fragrant and pretty flowers. The little plant, with its shining evergreen foliage, flourishes abundantly among decaying leaves in sandy soil, and puts forth its dainty blossoms late in June. It is one of the latest of the fragile wood-flowers which are so charming in the earlier year, and which have already begun to surrender in favor of their hardier, more self-assertive brethren of the fields and roadsides. The common name, pipsissewa, is evidently of Indian origin, and perhaps refers to the strengthening properties which the red men ascribed to it.

SPOTTED PIPSISSEWA.

Chimaphila maculata. Heath Family.

The spotted pipsissewa blossoms a little later than its twin-sister. Its slightly toothed leaves are conspicuously marked with white.

WHITE DAISY. WHITE-WEED. OX-EYED DAISY.

Chrysanthemum Leucanthemum. Composite Family (p. 13).

The common white daisy stars the June meadows with those gold-centred blossoms which delight the eyes of the beauty-

PLATE XIX

PIPSISSEWA.—*C. umbellata.*

69

lover while they make sore the heart of the farmer, for the "white-weed," as he calls it, is hurtful to pasture land and difficult to eradicate.

The true daisy is the *Bellis perennis* of England,—the

> Wee, modest crimson-tippit flower

of Burns. This was first called "day's eye," because it closed at night and opened at dawn,—

> That well by reason men it call may,
> The Daisie, or else the eye of the day,

sang Chaucer nearly four hundred years ago. In England our flower is called "ox-eye" and "moon daisy;" in Scotland, "dog-daisy."

The plant is not native to this country, but was brought from the Old World by the early colonists.

DAISY FLEABANE. SWEET SCABIOUS.

Erigeron annuus. Composite Family (p. 13).

Stem.—Stout, from three to five feet high, branched, hairy. *Leaves.*—Coarsely and sharply toothed, the lowest ovate, the upper narrower. *Flower-heads.*—Small, clustered, composed of both ray and disk-flowers, the former white, purplish, or pinkish, the latter yellow.

During the summer months the fields and waysides are whitened with these very common flowers which look somewhat like small white daisies or asters.

Another common species is *E. strigosus,* a smaller plant, with smaller flower-heads also, but with the white ray-flowers longer. The generic name is from two Greek words signifying *spring* and *an old man,* in allusion to the hoariness of certain species which flower in the spring. The fleabanes were so named from the belief that when burned they were objectionable to insects. They were formerly hung in country cottages for the purpose of excluding such unpleasant intruders.

MAYWEED. CHAMOMILE.

Anthemis Cotula. Composite Family (p. 13).

Stem.—Branching. *Leaves.*—Finely dissected. *Flower-heads.*—Composed of white ray and yellow disk-flowers, resembling the common white daisy.

In midsummer the pretty daisy-like blossoms of this strong-scented plant are massed along the roadsides. So nearly a counterpart of the common daisy do they appear that they are constantly mistaken for that flower. The smaller heads, with the yellow disk-flowers crowded upon a receptacle which is much more conical than that of the daisy, and the finely dissected, feathery leaves, serve to identify the Mayweed. The country-folk brew "chamomile tea" from these leaves, and through their agency raise painfully effective blisters in an emergency.

NEW JERSEY TEA. RED-ROOT.

Ceanothus Americanus. Buckthorn Family.

Root.—Dark red. *Stem.*—Shrubby, one to three feet high. *Flowers.*—White, small, clustered. *Calyx.*—White, petal-like, five-lobed, incurved. *Corolla.*—With five long-clawed hooded petals. *Stamens.*—Five. *Pistil.*—One, with three stigmas.

This shrubby plant is very common in dry woods. In July its white feathery flower-clusters brighten many a shady nook in an otherwise flowerless neighborhood. During the Revolution its leaves were used as a substitute for tea.

BASTARD TOADFLAX.

Comandra umbellata. Sandalwood Family.

Stem.—Eight to ten inches high, branching, leafy. *Leaves.*—Alternate, oblong, pale. *Flowers.*—Greenish-white, small, clustered. *Calyx.*—Bell or urn-shaped. *Corolla.*—None. *Stamens.*—Five. *Pistil.*—One.

The bastard toadflax is commonly found in dry ground, flowering in May or June. Its root forms parasitic attachments to the roots of trees.

WINTERGREEN. CHECKERBERRY. MOUNTAIN TEA.

Gaultheria procumbens. Heath Family.

Stem.—Three to six inches high, slender, leafy at the summit. *Leaves.*—Oval, shining, evergreen. *Flowers.*—White, growing from the axils of the leaves. *Calyx.*—Five-lobed. *Corolla.*—Urn-shaped, with five small teeth. *Stamens.*—Ten. *Pistil.*—One. *Fruit.*—A globular red berry.

He who seeks the cool shade of the evergreens on a hot July day is likely to discover the nodding wax-like flowers of this little plant. They are delicate and pretty, with a background of shining leaves. These leaves when young have a pleasant aromatic flavor similar to that of the sweet birch; they are sometimes used as a substitute for tea. The bright red berries are also edible and savory, and are much appreciated by the hungry birds and deer during the winter. If not thus consumed they remain upon the plant until the following spring when they either drop or rot upon the stem, thus allowing the seeds to escape.

WHITE SWEET CLOVER. WHITE MELILOT.

Melilotus alba. Pulse Family (p. 16).

Stem.—Two to four feet high. *Leaves.*—Divided into three-toothed leaflets. *Flowers.*—Papilionaceous, white, growing in spike-like racemes.

Like its yellow sister, *M. officinalis*, this plant is found blossoming along the roadsides throughout the summer. The flowers are said to serve as flavoring in Gruyère cheese, snuff, and smoking-tobacco, and to act like camphor when packed with furs to preserve them from moths, besides imparting a pleasant fragrance.

WATERLEAF.

Hydrophyllum Virginicum. Waterleaf Family.

One to two feet high. *Leaves.*—Divided into five to seven oblong, pointed, toothed divisions. *Flowers.*—White or purplish, in one-sided raceme-like clusters which are usually coiled from the apex when young. *Calyx.*—Five-parted. *Corolla.*—Five-cleft, bell-shaped. *Stamens.*—Five, protruding. *Pistil.*—One.

This plant is found flowering in summer in the rich woods.

PLATE XX

Fruit.

WINTERGREEN.—*G. procumbens.*

73

INDIAN PIPE. CORPSE-PLANT.

Monotropa uniflora. Heath Family.

A low fleshy herb from three to eight inches high, without green foliage, of a wax-like appearance, with colorless bracts in the place of leaves. *Flower.* —White or pinkish, single, terminal, nodding. *Calyx.* — Of two to four bract-like scales. *Corolla.*—Of four or five wedge-shaped petals. *Stamens.* —Eight or ten, with yellow anthers. *Pistil.*—One, with a disk-like, four or five-rayed stigma.

The effect of a cluster of these nodding, wax-like flowers in the deep woods of summer is singularly fairy-like. They spring from a ball of matted rootlets, and are parasitic, drawing their nourishment from decaying vegetable matter. In fruit the plant erects itself and loses its striking resemblance to a pipe. Its clammy touch, and its disposition to decompose and turn black when handled, has earned it the name of corpse-plant. It was used by the Indians as an eye-lotion, and is still believed by some to possess healing properties.

FIELD CHICKWEED.

Cerastium arvense. Pink Family.

Four to eight inches high. *Stems.*—Slender. *Leaves.*—Linear or narrowly lance-shaped. *Flowers.*—White, large, in terminal clusters. *Calyx.* —Usually of five sepals. *Corolla.*—Usually of five two-lobed petals which are more than twice the length of the calyx. *Stamens.*—Twice as many, or fewer than the petals. *Pistil.*—One, with as many styles as there are sepals.

This is one of the most noticeable of the chickweeds. Its starry flowers are found in dry or rocky places, blossoming from May till July.

The common chickweed, which besets damp places everywhere, is *Stellaria media;* this is much used as food for songbirds.

The long-leaved stitchwort, *S. longifolia*, is a species which is common in grassy places, especially northward. It has linear leaves, unlike those of *S. media*, which are ovate or oblong.

PLATE XXI

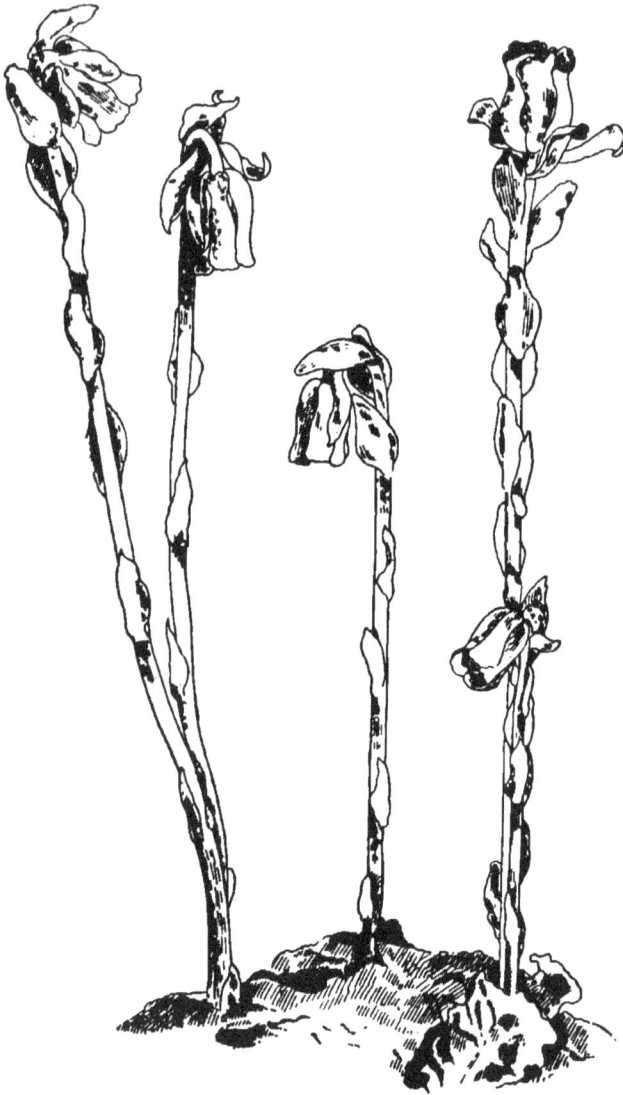

INDIAN PIPE.—*M. uniflora.*

ENCHANTER'S NIGHTSHADE.

Circæa Lutetiana. Evening Primrose Family.

Stem.—One or two feet high. *Leaves.*—Opposite, thin, ovate, slightly toothed. *Flowers.*—Dull white, small, growing in a raceme. *Calyx.*—Two-lobed. *Corolla.*—Of two petals. *Stamens.*—Two. *Pistil.*—One.

This insignificant and ordinarily uninteresting plant arrests attention by the frequency with which it is found flowering in the summer woods and along shady roadsides.

C. Alpina is a smaller, less common species, which is found along the mountains and in deep woods. Both species are burdened with the singularly inappropriate name of enchanter's nightshade. There is nothing in their appearance to suggest an enchanter or any of the nightshades. It seems, however, that the name of a plant called after the enchantress Circe, and described by Dioscorides nearly two thousand years ago, was accidentally transferred to this unpretentious genus.

THIMBLE-WEED.

Anemone Virginiana. Crowfoot Family.

Stem.—Two or three feet high. *Leaves.*—Twice or thrice cleft, the divisions again toothed or cleft. *Flowers.*—Greenish or sometimes white, borne on long, upright flower-stalks. *Calyx.*—Of five sepals. *Corolla.*—None. *Stamens and Pistils.*—Indefinite in number.

These greenish flowers, which may be found in the woods and meadows throughout the summer, are only striking by reason of their long, erect flower-stalks. The oblong, thimble-like fruit-head, which is rather noticeable in the later year, gives to the plant its common name.

CLEAVERS. GOOSE-GRASS. BEDSTRAW.

Galium Aparine. Madder Family.

Stem.—Weak and reclining, bristly. *Leaves.*— Lance-shaped, about eight in a whorl. *Flowers.*—White, small, growing from the axils of the leaves. *Calyx-teeth.* — Obsolete. *Corolla.* — Usually four-parted, wheel-shaped. *Stamens.*—Usually four. *Pistil.*—One with two styles. *Fruit.*—Globular, bristly, with hooked prickles.

This plant may be found in wooded or shady places throughout the continent. Its flowers, which appear in summer, are rather inconspicuous, one's attention being chiefly attracted by its many whorls of slender leaves.

BITTER-SWEET. WAX-WORK.

Celastrus scandens. Staff-tree Family.

Stem.—Woody, twining. *Leaves.*—Alternate, oblong, finely toothed, pointed. *Flowers.*—Small, greenish, or cream-color, in raceme-like clusters, appearing in June. *Pod.*—Orange-colored, globular, and berry-like, curling back in three divisions when ripe so as to display the scarlet covering of the seeds within.

The small flowers of the bitter-sweet, which appear in June, rarely attract attention. But in October no lover of color can fail to admire the deep orange pods which at last curl back so as advantageously to display the brilliant scarlet covering of the seeds. Perhaps we have no fruit which illuminates more vividly the roadside thicket of late autumn ; or touches with greater warmth those tumbled, overgrown walls which are so picturesque a feature in parts of the country, and do in a small way for our quiet landscapes what vine-covered ruins accomplish for the scenery of the Old World.

CULVER'S ROOT.

Veronica Virginica. Figwort Family.

Stem.—Straight and tall, from two to six feet high. *Leaves.*—Whorled, lance-shaped, finely toothed. *Flowers.*—White, small, growing in slender clustered spikes. *Calyx.*—Irregularly four or five-toothed. *Corolla.*—Four or five-lobed. *Stamens.*—Two, protruding. *Pistil.*—One.

The tall straight stems of the culver's root lift their slender spikes in midsummer to a height that seems strangely at variance with the habit of this genus. The small flowers, however, at once betray their kinship with the speedwells. Although it is, perhaps, a little late to look for the white wands of the black cohosh the two plants might easily be confused in the distance, as they have much the same aspect and seek alike the cool recesses of the woods. This same species grows in Japan and was introduced into English gardens nearly two hundred years ago. It is one of the many Indian remedies which were adopted by our forefathers.

BLACK COHOSH. BUGBANE. BLACK SNAKEROOT.

Cimicifuga racemosa. Crowfoot Family.

Stem.—Three to eight feet high. *Leaves.*—Divided, the leaflets toothed or incised. *Flowers.*—White, growing in elongated wand-like racemes. *Calyx.*—Of four or five white petal-like sepals, falling early. *Corolla.*—Of from one to eight white petals or transformed stamens. *Stamens.*—Numerous, with slender white filaments. *Pistils.*—One to three.

The tall white wands of the black cohosh shoot up in the shadowy woods of midsummer like so many ghosts. A curious-looking plant it is, bearing aloft the feathery flowers which have such an unpleasant odor that even the insects are supposed to avoid them. Fortunately they are sufficiently conspicuous to be admired at a distance, many a newly cleared hill-side and wood-border being lightened by their slender, torch-like racemes which flash upon us as we travel through the country. The plant was one of the many which the Indians believed to be efficacious for snake-bites. The generic name is from *cimex*—a bug, and *fugare* —to drive away.

COMMON ELDER.

Sambucus Canadensis. Honeysuckle Family.

Stems.—Scarcely woody, five to ten feet high. *Leaves.*—Divided into toothed leaflets. *Flowers.*—White, small, in flat-topped clusters. *Calyx.*— Lobes minute or none. *Corolla.*—With five spreading lobes. *Stamens.*— Five. *Pistil.*—One, with three stigmas. *Fruit.*—Dark-purple, berry-like.

The common elder borders the lanes and streams with its spreading flower-clusters in early summer, and in the later year is noticeable for the dark berries from which "elderberry wine" is brewed by the country people. The fine white wood is easily cut and is used for skewers and pegs. A decoction of the leaves serves the gardener a good purpose in protecting delicate plants from caterpillars. Evelyn wrote of it: "If the medicinal properties of the leaves, berries, bark, etc., were thoroughly known, I cannot tell what our countrymen could ail for which he might not fetch from every hedge, whether from sickness or wound."

The white pith can easily be removed from the stems, hence the old English name of bore-wood.

PLATE XXII

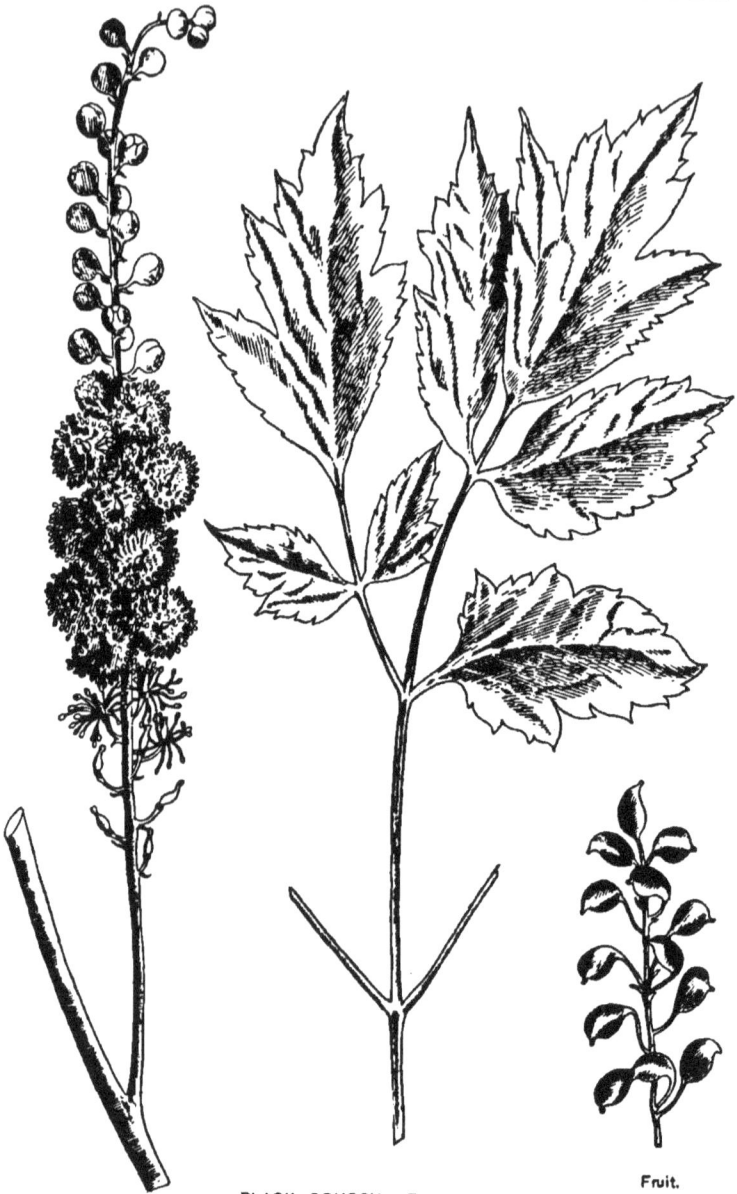

BLACK COHOSH.— *C. racemosa.*

Fruit.

The name elder is probably derived from the Anglo-Saxon *aeld*—a fire—and is thought to refer to the former use of the hollow branches in blowing up a fire.

SPURGE.

Euphorbia corollata. Spurge Family.

Stem.—Two or three feet high. *Leaves.*—Ovate, lance-shaped or linear. *Flowers.*—Clustered within the usually five-lobed, cup-shaped involucre which was formerly considered the flower itself ; the male flowers numerous and lining its base, consisting each of a single stamen ; the female flower solitary in the middle of the involucre, consisting of a three-lobed ovary with three styles, each style being two-cleft. *Pod.*—On a slender stalk, smooth.

In this plant the showy white appendages of the cup-shaped clustered involucres are usually taken for the petals of the flower ; only the botanist suspecting that the minute organs within these involucres really form a cluster of separate flowers of different sexes. While the most northerly range in the Eastern States of this spurge is usually considered to be New York, the botany states that it has been recently naturalized in Massachusetts. It blossoms from July till October.

PARTRIDGE VINE.

Mitchella repens. Madder Family.

Stems.— Smooth and trailing. *Leaves.*— Rounded, evergreen, veined with white. *Flowers.*—White, fragrant, in pairs. *Calyx.*—Four-toothed. *Corolla.*—Funnel-form, with four spreading lobes, bearded within. *Stamens.* —Four. *Pistil.*—One, its ovary united with that of its sister flower, its four stigmas linear.

At all times of the year this little plant faithfully fulfils its mission of adorning that small portion of the earth to which it finds itself rooted. But only the early summer finds the partridge vine exhaling its delicious fragrance from the delicate sister-blossoms which are its glory. Among the waxy flowers will be found as many of the bright red berries of the previous year as have been left unmolested by the hungry winter birds. This plant is found not only in the moist woods of North America,

PLATE XXIII

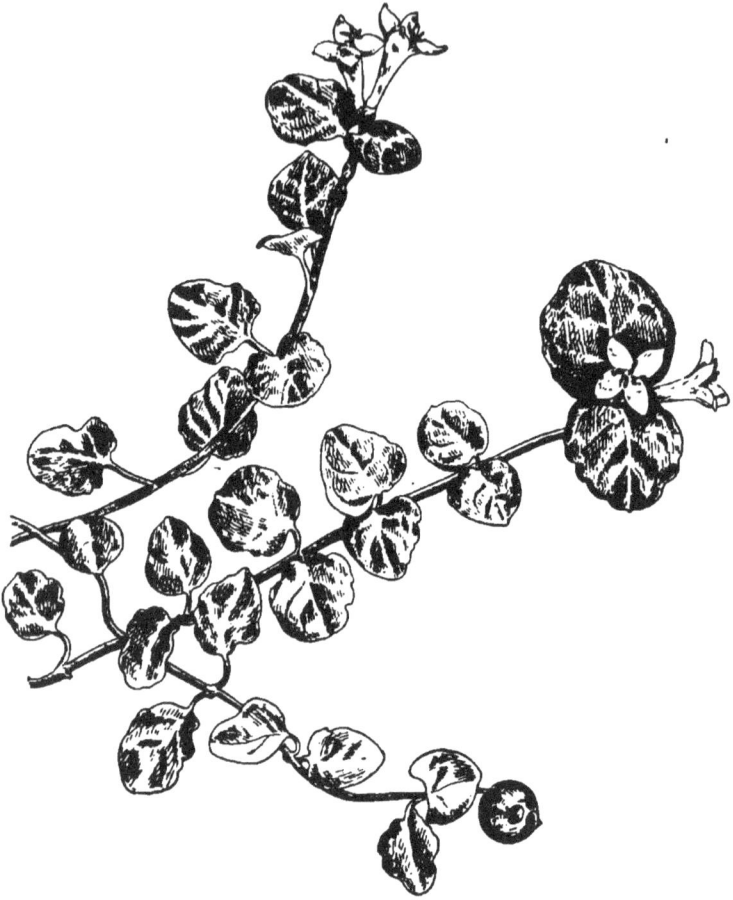

PARTRIDGE VINE.—*M. repens.*

but also in the forests of Mexico and Japan. It is a near relative of the dainty bluets or Quaker-ladies, and has the same peculiarity of dimorphous flowers (p. 232).

GREEN ORCHIS.

Habenaria virescens.

RAGGED FRINGED ORCHIS.

Habenaria lacera. Orchis Family (p. 17).

Leaves.—Oblong or lance-shaped. *Flowers.*—Greenish or yellowish-white, growing in a spike.

These two orchids are found in wet boggy places during the earlier summer, the green antedating the ragged fringed orchis by a week or more. The lip of the ragged fringed is three-parted, the divisions being deeply fringed, giving what is called in Sweet's " British Flower-Garden " an " elegantly jagged appearance." The lip of the green orchis is furnished with a tooth on each side and a strong protuberance in the middle. So far as superficial beauty and conspicuousness are concerned these flowers do scant justice to the brilliant family to which they belong, and equally excite the scornful exclamation, " You call *that* an orchid ! " when brought home for analysis or preservation.

BUTTON-BUSH.

Cephalanthus occidentalis. Madder Family.

A shrub three to eight feet high. *Leaves.*—Opposite or whorled in threes, somewhat oblong and pointed. *Flowers.*—Small, white, closely crowded in round button-like heads. *Calyx.*—Four-toothed. *Corolla.*—Four-toothed. *Stamens.*—Four. *Pistil.*—One, with a thread-like protruding style and blunt stigma.

This pretty shrub borders the streams and swamps throughout the country. Its button-like flower-clusters appear in midsummer. It belongs to the family of which the delicate bluet and fragrant partridge vine are also members. Its flowers have a jasmine-like fragrance.

MILD WATER-PEPPER.

Polygonum hydropiperoides. Buckwheat Family.

Stem. —One to three feet high, smooth, branching. *Leaves.*—Alternate, narrowly lance-shaped or oblong. *Flowers.*—White or flesh-color, small, growing in erect, slender spikes. *Calyx.*—Five-parted. *Corolla.*—None. *Stamens.*—Eight. *Pistil.*—One, usually with three styles.

These rather inconspicuous but very common flowers are found in moist places and shallow water.

The common knotweed, *P. aviculare,* which grows in such abundance in country door-yards and waste places, has slender, often prostrate, stems, and small greenish flowers, which are clustered in the axils of the leaves or spiked at the termination of the stems. This is perhaps the "hindering knotgrass" to which Shakespeare refers in the "Midsummer Night's Dream," so terming it, not on account of its knotted trailing stems, but because of the belief that it would hinder the growth of a child. In Beaumont and Fletcher's "Coxcomb" the same superstition is indicated:

> We want a boy
> Kept under for a year with milk and knotgrass.

It is said that many birds are nourished by the seeds of this plant.

CLIMBING FALSE BUCKWHEAT.

Polygonum scandens. Buckwheat Family.

Stem.—Smooth, twining, and climbing over bushes, eight to twelve feet high. *Leaves.*— Heart or arrow shaped, pointed, alternate. *Flowers.*— Greenish or pinkish, in racemes. *Calyx.*—Five-parted, with colored margins. *Corolla.*—None. *Stamens.*—Usually eight. *Pistil.*—One, with three stigmas. *Seed-vessel.*—Green, three-angled, winged, conspicuous in autumn.

In early summer this plant, which clambers so perseveringly over the moist thickets which line our country lanes, is comparatively inconspicuous. The racemes of small greenish flowers are not calculated to attract one's attention, and it is late summer or autumn before the thick clusters of greenish fruit composed of the winged seed-vessels arrest one's notice. At this time the vine is very beautiful and striking, and one wonders that it could have escaped detection in the earlier year.

Dalibarda repens. Rose Family.

Scape.—Low. *Leaves.*—Heart-shaped, wavy-toothed. *Flowers.*—White, one or two borne on each scape. *Calyx.*—Deeply five or six-parted, three of the divisions larger and toothed. *Corolla.*—Of five petals. *Stamens.*— Many. *Pistils.*—Five to ten.

The foliage of this pretty little plant suggests the violet ; while its white blossom betrays its kinship with the wild straw-berry. It may be found from June till August in woody places, being one of those flowers which we seek deliberately, whose charm is never decreased by its being thrust upon us inopportunely. Who can tell how much the attractiveness of the wild carrot, the dandelion, or butter-and-eggs would be enhanced were they so discreet as to withdraw from the common haunts of men into the shady exclusiveness which causes us to prize many far less beautiful flowers ?

STARRY CAMPION.

Silene stellata. Pink Family.

Stem.—Swollen at the joints, about three feet high. *Leaves.*—Whorled in fours, oval, taper-pointed. *Flowers.*—White, in a large pyramidal cluster. *Calyx.* — Inflated, five-toothed. *Corolla.* — Of five deeply fringed petals. *Stamens.*—Ten. *Pistil.*—One, with three styles.

In late July many of our wooded banks are decorated with the tall stems, whorled leaves, and prettily fringed flowers of the starry campion.

Closely allied to it is the bladder campion of the fields, *S. Cucubalus*, a much smaller plant, with opposite leaves, loosely clustered white flowers, a greatly inflated calyx, and two-cleft petals. This is an emigrant from Europe, which was first natu-ralized near Boston, and has now become wild in different parts of the country, quite overrunning some of the farm-lands which border the Hudson River.

PLATE XXiV

Dalibarda repens.

85

COLIC-ROOT. STAR-GRASS.

Aletris farinosa. Bloodwort Family.

Leaves.—Thin, lance-shaped, in a spreading cluster from the root. *Scape.*—Slender, two to three feet high. *Flowers.*—White, small, growing in a wand-like, spiked raceme. *Perianth.*—Six-cleft at the summit, oblong-tubular. *Stamens.*—Six. *Pistil.*—One, with style three-cleft at apex.

In summer we find these flowers in the grassy woods. The generic title is the Greek word for " a female slave who grinds corn," and refers to the mealy appearance of the blossoms.

TALL MEADOW RUE.

Thalictrum polygamum. Crowfoot Family.

Four to eight feet high. *Leaves.*—Divided into many firm, rounded leaflets. *Flowers.*—White, in large clusters ; some perfect, others unisexual. *Calyx.*—Of four or five small petal-like sepals which usually fall off very early. *Corolla.*—None. *Stamens.*—Numerous. *Pistils.*—Four to fifteen.

Where a stream trails its sluggish length through the fields of midsummer, its way is oftentimes marked by the tall meadow rue, the feathery, graceful flower-clusters of which erect themselves serenely above the myriad blossoms which are making radiant the wet meadows at this season. For here, too, we may search for the purple flag and fringed orchis, the yellow meadow lily, the pink swamp milkweed, each charming in its way, but none with the cool chaste beauty of the meadow rue. The staminate flowers of this plant are especially delicate and feathery.

WHITE AVENS.

Geum album. Rose Family.

Stem.—Slender, about two feet high. *Root-leaves.*—Divided into from three to five leaflets, or entire. *Stem-leaves.*—Three-lobed or divided, or only toothed. *Flowers.*—White. *Calyx.*—Deeply five-cleft, usually with five small bractlets alternating with its lobes. *Corolla.*—Of five petals. *Stamens.*—Numerous. *Pistils.*—Numerous, with hooked styles which become elongated in fruit.

The white avens is one of the less noticeable plants which border the summer woods, blossoming from May till August. Later the hooked seeds which grow in round burr-like heads

PLATE XXV

TALL MEADOW RUE.—*T. polygamum.*

secure wide dispersion by attaching themselves to animals or clothing. Other species of avens have more conspicuous golden-yellow flowers.

MEADOW-SWEET.

Spiræa salicifolia. Rose Family.

Stem.—Nearly smooth, two or three feet high. *Leaves.*—Alternate, somewhat lance-shaped, toothed. *Flowers.*—Small, white or flesh-color, in pyramidal clusters. *Calyx.*—Five-cleft. *Corolla.*—Of five rounded petals. *Stamens.*—Numerous. *Pistils.*—Five to eight.

The feathery spires of the meadow-sweet soar upward from the river banks and low meadows in late July. Unlike its pink sister, the steeple-bush, its leaves and stems are fairly smooth. The lack of fragrance in the flowers is disappointing, because of the hopes raised by the plant's common name. This is said by Dr. Prior to be a corruption of the Anglo-Saxon *mead-wort,* which signifies *honey-wine herb,* alluding to a fact which is mentioned in Hill's "Herbal," that "the flowers mixed with mead give it the flavor of the Greek wines."

Although the significance of many of the plant-names seems clear enough at first sight, such an example as this serves to show how really obscure it often is.

WHITE WATER-LILY.

Nymphæa odorata. Water-lily Family.

Leaves.—Rounded, somewhat heart-shaped, floating on the surface of the water. *Flowers.*—Large, white, or sometimes pink, fragrant. *Calyx.*—Of four sepals which are green without. *Corolla.*—Of many petals. *Stamens.*—Indefinite in number. *Pistil.*—With a many-celled ovary whose summit is tipped with a globular projection around which are the radiating stigmas.

This exquisite flower calls for little description. Many of us are so fortunate as to hold in our memories golden mornings devoted to its quest. We can hardly take the shortest railway journey in summer without passing some shadowy pool whose greatest adornment is this spotless and queenly blossom. The breath of the lily-pond is brought even into the heart of our cit-

PLATE XXVI

MEADOW-SWEET.—*S. salicifolia.*

ies where dark-eyed little Italians peddle clusters of the long-stemmed fragrant flowers about the streets.

In the water-lily may be seen an example of so-called *plant-metamorphosis.* The petals appear to pass gradually into stamens, it being difficult to decide where the petals end and the stamens begin. But whether stamens are transformed petals, or petals transformed stamens seems to be a mooted question. In Gray we read, " Petals numerous, in many rows, the innermost gradually passing into stamens," while Mr. Grant Allen writes : " Petals are in all probability enlarged and flattened stamens, which have been set apart for the work of attracting insects," and goes on to say, " Flowers can and do exist without petals, . . . but no flower can possibly exist without stamens, which are one of the two essential reproductive organs in the plant." From this he argues that it is more rational to consider a petal a transformed stamen than *vice versa.* To go further into the subject here would be impossible, but a careful study of the water-lily is likely to excite one's curiosity in the matter.

WHITE VERVAIN.

Verbena urticæfolia. Verbena Family.

Three to five feet high. *Leaves.*—Oval, coarsely toothed. *Flowers.*—Small, white, in slender spikes, otherwise resembling Purple Vervain.

It almost excites one's incredulity to be told that this uninteresting looking plant, which grows rankly along the highways, is an importation from the tropics, yet for this statement the botany is responsible.

ROUND-LEAVED SUNDEW.

Drosera rotundifolia. Sundew Family.

Scape.—A few inches high. *Leaves.*—Rounded, abruptly narrowed into spreading, hairy leaf-stalks ; beset with reddish, gland-bearing bristles. *Flowers.*—White, growing in a one-sided raceme, which so nods at its apex that the fresh-blown blossom is always uppermost. *Calyx.*—Of five sepals.

Corolla.—Of five petals. *Pistil.*—One, with three or five styles, which are sometimes so deeply two-parted as to be taken for twice as many.

What's this I hear
About the new carnivora ?
Can little plants
Eat bugs and ants
And gnats and flies ?
A sort of retrograding :
Surely the fare
Of flowers is air,
Or sunshine sweet ;
They shouldn't eat,
Or do aught so degrading !

But by degrees we are learning to reconcile ourselves to the fact that the more we study the plants the less we are able to attribute to them altogether unfamiliar and ethereal habits. We find that the laws which control their being are strangely suggestive of those which regulate ours, and after the disappearance of the shock which attends the shattered illusion, their charm is only increased by the new sense of kinship.

The round-leaved sundew is found blossoming in many of our marshes in midsummer. When the sun shines upon its leaves they look as though covered with sparkling dewdrops, hence its common name. These drops are a glutinous exudation, by means of which insects visiting the plant are first captured ; the reddish bristles then close tightly about them, and it is supposed that their juices are absorbed by the plant. At all events the rash visitor rarely escapes. In many localities it is easy to secure any number of these little plants and to try for one's self the rather grewsome experiment of feeding them with small insects. Should the tender-hearted recoil from such reckless slaughter, they might confine their offerings on the altar of science to mosquitoes, small spiders, and other deservedly unpopular creatures.

D. Americana is a very similar species, with longer, narrower leaves.

The thread-leaved sundew, *D. filiformis* has fine, thread-like leaves and pink flowers, and is found in wet sand along the coast.

POKEWEED. GARGET. PIGEON-BERRY.

Phytolacca decandra. Pokeweed Family.

Stems.—In length from six to ten feet high; purple-pink or bright red, stout. *Leaves.*—Large, alternate, veiny. *Flowers.*—White or pinkish, the green ovaries conspicuous, growing in racemes. *Calyx.*—Of five rounded or petal-like sepals, pinkish without. *Corolla.*—None. *Stamens.*—Ten. *Pistil.*—One, with ten styles. *Fruit.*—A dark purplish berry.

There is a vigor about this native plant which is very pleasing. In July it is possible that we barely notice the white flowers and large leaves; but when in September the tall purple stems rear themselves above their neighbors in the roadside thicket, the leaves look as though stained with wine, and the long clusters of rich dark berries hang heavily from the branches, we cannot but admire its independent beauty. The berries serve as food for the birds. A tincture of them at one time acquired some reputation as a remedy for rheumatism. In Pennsylvania they have been used with whiskey to make a so-called "port-wine." From their dark juice arose the name of "red-ink plant," which is common in some places. The large roots are poisonous, but the acrid young shoots are rendered harmless by boiling, and are eaten like asparagus, being quite as good, I have been told by country people.

Despite the difference in the spelling of the names, it has been suggested that the plant was called after President Polk. This is most improbable, as it was common throughout the country long before his birth, and its twigs are said to have been plucked and worn by his followers during his campaign for the Presidency.

WHITE FRINGED ORCHIS.

Habenaria blephariglottis. Orchis Family (p. 17).

About one foot high. *Leaves.*—Oblong or lance-shaped, the upper passing into pointed bracts. *Flowers.*—Pure white, with a slender spur and fringed lip; growing in an oblong spike.

This seems to me the most exquisite of our native orchids. The fringed lips give the snowy, delicate flowers a feathery appearance as they gleam from the shadowy woods of midsummer,

PLATE XXVII

Fruit.

POKEWEED.—*P. decandra.*

93

or from the peat-bogs where they thrive best; or perhaps they spire upward from among the dark green rushes which border some lonely mountain lake. Like the yellow fringed orchis (Pl. LII), which they greatly resemble in general structure, they may be sought in vain for many seasons and then will be discovered one midsummer day lavishing their spotless loveliness upon some unsuspected marsh which has chanced to escape our vigilance.

RATTLESNAKE-PLANTAIN.

Goodyera pubescens. Orchis Family (p. 17).

Scape.—Six to twelve inches high. *Leaves.*—From the root in a sort of flat rosette; conspicuously veined with white; thickish, evergreen. *Flowers.* —Small, greenish-white, crowded in a close spike.

The flowers of the rattlesnake-plantain appear in late summer and are less conspicuous than the prettily tufted, white-veined leaves which may be found in the rich woods throughout the year. The plant has been reputed an infallible cure for hydrophobia and snake-bites. It is said that the Indians had such faith in its remedial virtues that they would allow a snake to drive its fangs into them for a small sum, if they had these leaves on hand to apply to the wound.

COMMON YARROW. MILFOIL.

Achillea Millefolium. Composite Family (p. 13).

Stem.—Simple at first, often branching near the summit. *Leaves.*— Divided into finely toothed segments. *Flower-heads.*—White, occasionally pink, clustered, small, made up of both ray and disk-flowers.

This is one of our most frequent roadside weeds, blossoming throughout the summer and late into the autumn. Tradition claims that it was used by Achilles to cure the wounds of his soldiers, and the genus is named after that mighty hero. It still forms one of the ingredients of an ointment valued by the Scotch Highlanders. The early English botanists called the plant "nosebleed," "because the leaves being put into the nose caused it to bleed;" and Gerarde writes that "Most men say that the leaves chewed, and especially greene, are a remedie for the toothache."

PLATE XXVIII

WILD CARROT.—*D. carota.*   YARROW. — *A. millefolium.*

These same pungent leaves also won it the name of "old man's pepper," while in Sweden its title signifies *field hop*, and refers to its employment in the manufacture of beer. Linnæus considered the beer thus brewed to be more intoxicating than that in which hops were utilized. The old women of the Orkney Islands hold "milfoil tea" in high repute, believing it to be gifted with the power of dispelling melancholy. In Switzerland a good vinegar is said to be made from the Alpine species. The plant is cultivated in the gardens of Madeira, where so many beautiful, and in our eyes rare, flowers grow in wild profusion.

WILD CARROT. BIRD'S NEST. QUEEN ANNE'S LACE.

Daucus carota. Parsley Family (p. 15).

Stems.—Tall and slender. *Leaves.*—Finely dissected. *Flowers.*— White, in a compound umbel, forming a circular flat-topped cluster.

When the delicate flowers of the wild carrot are still unsoiled by the dust from the highway, and fresh from the early summer rains, they are very beautiful, adding much to the appearance of the roadsides and fields along which they grow so abundantly as to strike despair into the heart of the farmer, for this is, perhaps, the "peskiest" of all the weeds with which he has to contend. As time goes on the blossoms begin to have a careworn look and lose something of the cobwebby aspect which won them the title of Queen Anne's lace. In late summer the flower-stalks erect themselves, forming a concave cluster which has the appearance of a bird's nest. I have read that a species of bee makes use of this ready-made home, but have never seen any indications of such an occupancy.

This is believed to be the stock from which the garden carrot was raised. The vegetable was well known to the ancients, and we learn from Pliny that the finest specimens were brought to Rome from Candia. When it was first introduced into Great Britain is not known, although the supposition is that it was brought over by the Dutch during the reign of Elizabeth. In the writings of Parkinson we read that the ladies wore carrot-leaves in their hair in place of feathers. One can picture the dejected appearance of a ball-room belle at the close of an entertainment.

WATER HEMLOCK. SPOTTED COWBANE.

Cicuta maculata. Parsley Family (p. 15).

Stem.—Smooth, stout, from two to six feet high, streaked with purple. *Leaves.*—Twice or thrice-compound, leaflets coarsely toothed. *Flowers.*— White, in compound umbels, the little umbels composed of numerous flowers.

This plant is often confused with the wild carrot, the sweet Cicely, and other white-flowered members of the Parsley family ; but it can usually be identified by its purple-streaked stem. The umbels of the water-hemlock are also more loosely clustered than those of the carrot, and their stalks are much more unequal. It is commonly found in marshy ground, blossoming in midsummer. Its popular names refer to its poisonous properties, its root being said to contain the most dangerous vegetable-poison native to our country and to have been frequently confounded with that of the edible sweet Cicely with fatal results.

MOCK BISHOP-WEED.

Discopleura capillacea. Parsley Family (p. 15).

One or two feet high, occasionally much taller. *Stems.*—Branching. *Leaves.*—Dissected into fine, thread-like divisions. *Flowers.*—White, very small, growing in compound umbels with thread-like bracts.

This plant blossoms all summer in wet meadows, both inland and along the coast ; but it is especially common in the salt-marshes near New York City. It probably owes its English name to the fancied resemblance between the bracted flower-clusters and a bishop's cap. Its effect is feathery and delicate.

SWEET CICELY.

Osmorrhiza longistylis. Parsley Family (p. 15).

One to three feet high. *Root.*—Thick, aromatic, edible. *Leaves.*— Twice or thrice-compound. *Flowers.*—White, growing in a few-rayed compound umbel.

The roots of the sweet Cicely are prized by country children for their pleasant flavor. Great care should be taken not to confound this plant with the water-hemlock, which is very poisonous, and which it greatly resembles, although flowering earlier in the year. The generic name is from two Greek words which signify *scent* and *root*.

WATER-PARSNIP.

Sium cicutafolium. Parsley Family (p. 15).

Two to six feet high. *Stem.*—Stout. *Leaves.*—Divided into from three to eight pairs of sharply toothed leaflets. *Flowers.*—White, in compound umbels.

This plant is found growing in water or wet places throughout North America.

ARROW-HEAD.

Sagittaria variabilis. Water-plantain Family.

Scape.—A few inches to several feet high. *Leaves.*—Arrow-shaped. *Flowers.*—White, unisexual, in whorls of three on the leafless scape. *Calyx.* —Of three sepals. *Corolla.*—Of three white, rounded petals. *Stamens and Pistils.*—Indefinite in number, occurring in different flowers, the lower whorls of flowers usually being pistillate, the upper staminate.

Among our water-flowers none are more delicately lovely than those of the arrow-head. Fortunately the ugly and inconspicuous female flowers grow on the lower whorls, while the male ones, with their snowy petals and golden centres, are arranged about the upper part of the scape, where the eye first falls. It is a pleasure to chance upon a slow stream whose margins are bordered with these fragile blossoms and bright, arrow-shaped leaves.

WATER-PLANTAIN.

Alisma Plantago. Water-plantain Family.

Scape.—One to three feet high, bearing the flowers in whorled, panicled branches. *Leaves.*—From the root, oblong, lance-shaped or linear, mostly rounded or heart-shaped at base. *Flowers.*—White or pale pink, small, in large, loose clusters which branch from the scape. *Calyx.*—Of three sepals. *Corolla.*—Of three petals. *Stamens.*—Usually six. *Pistils.*—Many, on a flattened receptacle.

The water-plantain is nearly related to the arrow-head, and is often found blossoming with it in marshy places or shallow water.

PLATE XXIX

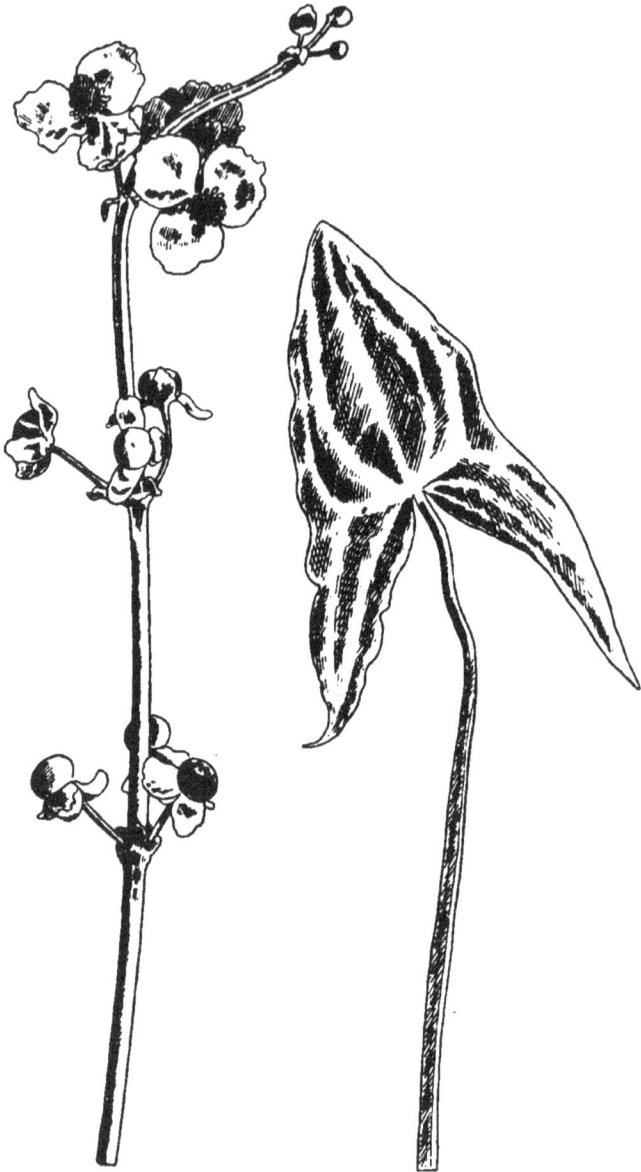

ARROW-HEAD.—*S. variabilis.*

GROUND CHERRY.

Physalis Virginiana. Nightshade Family.

A strong-scented, low, much-branched and spreading herb. *Leaves.*—Somewhat oblong or heart-shaped, wavy-toothed. *Flowers.*—Greenish or yellowish-white, solitary on nodding flower-stalks. *Calyx.*—Five-cleft ; enlarging and much inflated in fruit, loosely enclosing the berry. *Corolla.*—Between wheel-shaped and funnel-form. *Stamens.*—Five, erect, with yellow anthers. *Pistil.*—One. *Fruit.*—A green or yellow edible berry which is loosely enveloped in the much-inflated calyx.

We find the ground cherry in light sandy soil, and are more apt to notice the loosely enveloped berry of the late year than the rather inconspicuous flowers which appear in summer.

TURTLE-HEAD.

Chelone glabra. Figwort Family.

One to seven feet high. *Stem.*—Smooth, upright, branching. *Leaves.*—Opposite, lance-shaped, toothed. *Flowers.*—White or pinkish, growing in a spike or close cluster. *Calyx.*—Of five sepals. *Corolla.*—Two-lipped, the upper lip broad and arched, notched at the apex, lower lip three-lobed at the apex, woolly bearded in the throat. *Stamens.*—Four perfect ones, with woolly filaments and very woolly, heart-shaped anthers, and one small sterile one. *Pistil.*—One.

It seems to have been my fate to find the flowers which the botany relegates to " dry, sandy soil " flourishing luxuriantly in marshes ; and to encounter the flowers which by rights belong to " wet woods " flaunting themselves in sunny meadows. This cannot be attributed to the natural depravity of inanimate objects, for what is more full of life than the flowers ?—and no one would believe in their depravity except perhaps the amateur-botanist who is endeavoring to master the different species of golden-rods and asters. Therefore it is pleasant to record that I do not remember ever having met a turtle-head, which is assigned by the botany to " wet places," which had not gotten as close to a stream or a marsh or a moist ditch as it well could without actually wetting its feet. The flowers of this plant are more odd and striking than pretty. Their appearance is such that their common name seems fairly appropriate. I have heard unbotanical people call them " white closed gentians."

PLATE XXX

TURTLE-HEAD.—*C. glabra.*

COMMON DODDER. LOVE VINE.

Cuscuta Gronovii. Convolvulus Family.

Stems.—Yellow or reddish, thread-like, twining, leafless. *Flowers.*—White, in close clusters. *Calyx.*—Five-cleft. *Corolla.*—With five spreading lobes. *Stamens.*—Five. *Pistil.*—One, with two styles.

Late in the summer we are perhaps tempted deep into some thicket by the jasmine-scented heads of the button-bush or the fragrant spikes of the clethra, and note for the first time the tangled golden threads and close white flower-clusters of the dodder. If we try to trace to their source these twisted stems, which the Creoles know as "angels' hair," we discover that they are fastened to the bark of the shrub or plant about which they are twining by means of small suckers ; but nowhere can we find any connection with the earth, all their nourishment being extracted from the plant to which they are adhering. Originally this curious herb sprang from the ground which succored it until it succeeded in attaching itself to some plant ; having accomplished this it severed all connection with mother-earth by the withering away or snapping off of the stem below.

The flax-dodder, *C. Epilinum*, is a very injurious plant in European flax-fields. It has been sparingly introduced into this country with flax-seed.

TRAVELLER'S JOY. VIRGIN'S BOWER.

Clematis Virginiana. Crowfoot Family.

Stem.—Climbing, somewhat woody. *Leaves.*—Opposite, three-divided. *Flowers.*—Whitish, in clusters, unisexual. *Calyx.*—Of four petal-like sepals. *Corolla.*—None. *Stamens and Pistils.*—Indefinite in number, occurring on different plants.

In July and August this beautiful plant, covered with its white blossoms and clambering over the shrubs which border the country lanes, makes indeed a fitting bower for any maid or traveller who may chance to be seeking shelter. Later in the year the seeds with their silvery plumes give a feathery effect which is very striking.

This graceful climber works its way by means of its bending

PLATE XXXI

Fruit-cluster.

TRAVELLER'S JOY.—*Clematis Virginiana.*

103

or clasping leaf-stalks. Darwin has made interesting experiments regarding the movements of the young shoots of the *Clematis*. He discovered that, "one revolved describing a broad oval, in five hours, thirty minutes; and another in six hours, twelve minutes; they follow the course of the sun."

SWEET PEPPERBUSH. WHITE ALDER.

Clethra alnifolia. Heath Family.

A shrub from three to ten feet high. *Leaves.*—Alternate, ovate, sharply toothed. *Flowers.*—White, growing in clustered finger-like racemes. *Calyx.*—Of five sepals. *Corolla.*—Of five oblong petals. *Stamens.*—Ten, protruding. *Pistil.*—One, three-cleft at apex.

Nearly all our flowering shrubs are past their glory by mid-summer, when the fragrant blossoms of the sweet pepperbush begin to exhale their perfume from the cool thickets which line the lanes along the New England coast. There is a certain luxuriance in the vegetation of this part of the country in August which is generally lacking farther inland, where the fairer flowers have passed away, and the country begins to show the effects of the long days of heat and drought. The moisture of the air, and the peculiar character of the soil near the sea, are responsible for the freshness and beauty of many of the late flowers which we find in such a locality.

Clethra is the ancient Greek name for the alder, which this plant somewhat resembles in foliage.

THORN-APPLE. JAMESTOWN WEED.

Datura Stramonium. Nightshade Family.

Stem.—Smooth and branching. *Leaves.*—Ovate, wavy-toothed or angled. *Flowers.*—White, large and showy, on short flower-stalks from the forks of the branching stem. *Calyx.*—Five-toothed. *Corolla.*—Funnel-form, the border five-toothed. *Stamens.* — Five. *Pistil.* — One. *Fruit.* — Green, globular, prickly.

The showy white flowers of the thorn-apple are found in waste places during the summer and autumn, a heap of rubbish forming their usual unattractive background. The plant is a rank, ill-scented one, which was introduced into our country from Asia.

It was so associated with civilization as to be called the "white man's plant" by the Indians.

Its purple-flowered relative, *D. Tatula*, is an emigrant from the tropics. This genus possesses narcotic-poisonous properties.

WILD BALSAM-APPLE.

Echinocystis lobata. Gourd Family.

Stem.—Climbing, nearly smooth, with three-forked tendrils. *Leaves.*—Deeply and sharply five-lobed. *Flowers.*—Numerous, small, greenish-white, unisexual ; the staminate ones growing in long racemes, the pistillate ones in small clusters or solitary. *Fruit.*—Fleshy, oval, green, about two inches long, clothed with weak prickles.

This is an ornamental climber which is found bearing its flowers and fruit at the same time. It grows in rich soil along rivers in parts of New England, Pennsylvania, and westward ; and is often cultivated in gardens, making an effective arbor-vine. The generic name is from two Greek words which signify *hedgehog* and *bladder*, in reference to the prickly fruit.

WHITE ASTERS.

Aster. Composite Family (p. 13).

Flower-heads.—Composed of white ray-flowers with a centre of yellow disk-flowers.

While we have far fewer species of white than of blue or purple asters, some of these few are so abundant in individuals as to hold their own fairly well against their bright-hued rivals.

The slender zig-zag stems, thin, coarsely toothed, heart-shaped leaves, and white, loosely clustered flower-heads of *A. corymbosus*, are noticeable along the shaded roadsides and in the open woods of August.

Bordering the dry fields at this same season are the spreading wand-like branches, thickly covered with the tiny flower-heads as with snowflakes, of *A. ericoides*.

A. umbellatus is the tall white aster of the swamps and moist thickets. It sometimes reaches a height of seven feet, and can be identified by its long tapering leaves and large, flat flower-clusters.

A beautiful and abundant seaside species is *A. multiflorus.* Its small flower-heads are closely crowded on the low, bushy, spreading branches ; its leaves are narrow, rigid, crowded, and somewhat hoary. The whole effect of the plant is heath-like ; it also somewhat suggests an evergreen.

BONESET. THOROUGHWORT.

Eupatorium perfoliatum. Composite Family (p. 13).

Stem.—Stout and hairy, two to four feet high. *Leaves.*—Opposite, widely spreading, lance-shaped, united at the base around the stem. *Flower-heads.*—Dull white, small, composed entirely of tubular blossoms borne in large clusters.

To one whose childhood was passed in the country some fifty years ago the name or sight of this plant is fraught with unpleasant memories. The attic or wood-shed was hung with bunches of the dried herb which served as so many grewsome warnings against wet feet, or any over-exposure which might result in cold or malaria. A certain Nemesis, in the shape of a nauseous draught which was poured down the throat under the name of "boneset tea," attended such a catastrophe. The Indians first discovered its virtues, and named the plant ague-weed. Possibly this is one of the few herbs whose efficacy has not been over-rated. Dr. Millspaugh says : " It is prominently adapted to cure a disease peculiar to the South, known as break-bone fever (Dengue), and it is without doubt from this property that the name boneset was derived."

WHITE SNAKEROOT.

Eupatorium ageratoides. Composite Family (p. 13).

About three feet high. *Stem.*—Smooth and branching. *Leaves.*—Opposite, long-stalked, broadly ovate, coarsely and sharply toothed. *Flower-heads.*—White, clustered, composed of tubular blossoms.

Although this species is less common than boneset, it is frequently found blossoming in the rich Northern woods of late summer.

PLATE XXXII

BONESET.—*E. perfoliatum.*

107

CLIMBING HEMP-WEED.

Mikania scandens. Composite Family (p. 13).

Stem.—Twining and climbing, nearly smooth. *Leaves.*—Opposite, somewhat triangular-heart-shaped, pointed, toothed at the base. *Flower-heads.*—Dull white or flesh-color, composed of four tubular flowers ; clustered, resembling boneset.

In late summer one often finds the thickets which line the slow streams nearly covered with the dull white flowers of the climbing hemp-weed. At first sight the likeness to the boneset is so marked that the two plants are often confused, but a second glance discovers the climbing stems and triangular leaves which clearly distinguish this genus.

LADIES' TRESSES.

Spiranthes cernua. Orchis Family (p. 17).

Stem.—Leafy below, leafy-bracted above, six to twenty inches high. *Leaves.*— Linear-lance-shaped, the lowest elongated. *Flowers.*—White, fragrant, the lips wavy or crisped ; growing in slender spikes.

This pretty little orchid is found in great abundance in September and October. The botany relegates it to "wet places," but I have seen dry upland pastures as well as low-lying swamps profusely flecked with its slender, fragrant spikes. The braided appearance of these spikes would easily account for the popular name of ladies' tresses ; but we learn that the plant's English name was formerly "ladies' *traces*," from a fancied resemblance between its twisted clusters and the lacings which played so important a part in the feminine toilet. I am told that in parts of New England the country people have christened the plant "wild hyacinth."

The flowers of *S. gracilis* are very small, and grow in a much more slender, one-sided spike than those of *S. cernua.* They are found in the dry woods and along the sandy hill sides from July onward.

PLATE XXXIII

LADIES' TRESSES.—*S. cernua.*

GREEN-FLOWERED MILKWEED.

Asclepias verticillata. Milkweed Family.

Stem.—Slender, very leafy to the summit. *Leaves.*—Very narrow, from three to six in a whorl. *Flowers.*—Greenish-white, in small clusters at the summit and along the sides of the stem. *Fruit.*—Two erect pods, one often stunted.

This species is one commonly found on dry uplands, especially southward, with flowers resembling in structure those of the other milkweeds. (Pl. .)

GROUNDSEL TREE.

Baccharis halimifolia. Composite Family (p. 13).

A shrub from six to twelve feet high. *Leaves.*—Somewhat ovate and wedge-shaped, coarsely toothed on the upper entire. *Flower-heads.*—Whitish or yellowish, composed of unisexual tubular flowers, the stamens and pistils occurring on different plants.

Some October day, as we pick our way through the salt marshes which lie back of the beach, we may spy in the distance a thicket which looks as though composed of such white-flowered shrubs as belong to June. Hastening to the spot we discover that the silky-tufted seeds of the female groundsel-tree are responsible for our surprise. The shrub is much more noticeable and effective at this season than when—a few weeks previous— it was covered with its small white or yellowish flower-heads.

GRASS OF PARNASSUS.

Parnassia Caroliniana. Saxifrage Family.

Stem.—Scape-like, nine inches to two feet high, with usually one small rounded leaf clasping it below ; bearing at its summit a single flower. *Leaves.*—Thickish, rounded, often heart-shaped, from the root. *Flower.*— White or cream-color, veiny. *Calyx.*—Of five slightly united sepals. *Corolla.*—Of five veiny petals. *True Stamens.*—Five, alternate with the petals, and with clusters of sterile gland-tipped filaments. *Pistil.*—One, with four stigmas.

Gerarde indignantly declares that this plant has been described by blind men, not "such as are blinde in their eyes, but in their understandings, for if this plant be a kind of grasse then

PLATE XXXIV

GRASS OF PARNASSUS.—*P. Caroliniana.*

III

may the Butter-burre or Colte's-foote be reckoned for grasses—as also all other plants whatsoever." But if it covered Parnassus with its delicate veiny blossoms as abundantly as it does some moist New England meadows each autumn, the ancients may have reasoned that a plant almost as common as grass must somehow partake of its nature. The slender-stemmed, creamy flowers are never seen to better advantage than when disputing with the fringed gentian the possession of some luxurious swamp.

PEARLY EVERLASTING.

Anaphilis margaritacea. Composite Family (p. 13).

Stem.—Erect, one or two feet high, leafy. *Leaves.*—Broadly linear to lance-shaped. *Flower-heads.*—Composed entirely of tubular flowers with very numerous pearly white involucral scales.

This species is common throughout our Northern woods and pastures, blossoming in August. Thoreau writes of it in September : " The pearly everlasting is an interesting white at present. Though the stems and leaves are still green, it is dry and unwithering like an artificial flower ; its white, flexuous stem and branches, too, like wire wound with cotton. Neither is there any scent to betray it. Its amaranthine quality is instead of high color. Its very brown centre now affects me as a fresh and original color. It monopolizes small circles in the midst of sweet fern, perchance, on a dry hill-side."

FRAGRANT LIFE-EVERLASTING.

Gnaphalium polycephalum. Composite Family (p. 13).

Stem.—Erect, one to three feet high, woolly. *Leaves.*—Lance-shaped. *Flower-heads.*—Yellowish-white, clustered at the summit of the branches, composed of many tubular flowers.

This is the "fragrant life-everlasting," as Thoreau calls it, of late summer. It abounds in rocky pastures and throughout the somewhat open woods.

NOTE.—Flowers so faintly tinged with color as to give a white effect in the mass or at a distance are placed in the White section : *greenish* or *greenish-white* flowers are also found here. The Moth Mullein (p. 152) and Bouncing Bet (p. 196) are found frequently bearing white flowers : indeed, white varieties of flowers which are usually colored, need never surprise one.

II

YELLOW

MARSH MARIGOLD.

Caltha palustris. Crowfoot Family.

Stem.—Hollow, furrowed. *Leaves.*—Rounded, somewhat kidney-shaped. *Flowers.*—Golden-yellow. *Calyx.*—Of five to nine petal-like sepals. *Corolla.*—None. *Stamens.*—Numerous. *Pistils.*—Five to ten, almost without styles.

> Hark, hark ! the lark at Heaven's gate sings,
> And Phœbus 'gins arise,
> His steeds to water at those springs,
> On chaliced flowers that lies :
> And winking Mary-buds begin
> To ope their golden eyes ;
> With everything that pretty is—
> My lady sweet, arise !
> Arise, arise.—*Cymbeline.*

We claim—and not without authority—that these " winking Mary-buds " are identical with the gay marsh marigolds which border our springs and gladden our wet meadows every April. There are those who assert that the poet had in mind the garden marigold—*Calendula*—but surely no cultivated flower could harmonize with the spirit of the song as do these gleaming swamp blossoms. We will yield to the garden if necessary—

> The marigold that goes to bed with the sun
> And with him rises weeping—

of the " Winter's Tale," but insist on retaining for that larger, lovelier garden in which we all feel a certain sense of possession —even if we are not taxed on real estate in any part of the country—the " golden eyes " of the Mary-buds, and we feel strengthened in our position by the statement in Mr. Robinson's " Wild Garden " that the marsh marigold is so abundant along certain English rivers as to cause the ground to look as though paved with gold at those seasons when they overflow their banks.

These flowers are peddled about our streets every spring under the name of cowslips—a title to which they have no claim, and which is the result of that reckless fashion of christening unrecognized flowers which is so prevalent, and which is responsible for so much confusion about their English names.

The derivation of marigold is somewhat obscure. In the " Grete Herball " of the sixteenth century the flower is spoken of as *Mary Gowles*, and by the early English poets as *gold* simply. As the first part of the word might be derived from the Anglo-Saxon *mere*—a marsh, it seems possible that the entire name may signify *marsh-gold*, which would be an appropriate and poetic title for this shining flower of the marshes.

SPICE-BUSH. BENJAMIN-BUSH. FEVER-BUSH.

Lindera Benzoin. Laurel Family.

An aromatic shrub from six to fifteen feet high. *Leaves.*—Oblong, pale underneath. *Flowers.*—Appearing before the leaves in March or April, honey-yellow, borne in clusters which are composed of smaller clusters, surrounded by an involucre of four early falling scales. *Fruit.*—Red, berry-like, somewhat pear-shaped.

These are among the very earliest blossoms to be found in the moist woods of spring. During the Revolution the powdered berries were used as a substitute for allspice ; while at the time of the Rebellion the leaves served as a substitute for tea.

YELLOW ADDER'S TONGUE. DOG'S TOOTH VIOLET.

Erythronium Americanum. Lily Family.

Scape.—Six to nine inches high, one-flowered. *Leaves.*—Two, oblong-lance-shaped, pale green mottled with purple and white. *Flower.*—Rather large, pale yellow marked with purple, nodding. *Perianth.*—Of six recurved or spreading sepals. *Stamens.*—Six. *Pistil.*—One.

The white blossoms of the shad-bush gleam from the thicket, and the sheltered hill-side is already starred with the blood-root and anemone when we go to seek the yellow adder's tongue. We direct our steps toward one of those hollows in the wood which is watered by such a clear gurgling brook as must appeal to every country-loving heart ; and there where the pale April sunlight filters through the leafless branches, nod myriads of

PLATE XXXV

MARSH MARIGOLD.—*C. palustris.*

115

these lilies, each one guarded by a pair of mottled, erect, sentinel-like leaves.

The two English names of this plant are unsatisfactory and inappropriate. If the marking of its leaves resembles the skin of an adder why name it after its tongue? And there is equally little reason for calling a lily a violet. Mr. Burroughs has suggested two pretty and significant names. "Fawn lily," he thinks, would be appropriate, because a fawn is also mottled, and because the two leaves stand up with the alert, startled look of a fawn's ears. The speckled foliage and perhaps its flowering season are indicated in the title "trout-lily," which has a spring-like flavor not without charm. It is said that the early settlers of Pennsylvania named the flower "yellow snowdrop," in memory of their own "harbinger of spring."

The white adder's tongue, *E. albidum*, is a species which is usually found somewhat westward.

CELANDINE.

Chelidonium majus. Poppy Family.

Stem.—Brittle, with saffron-colored, acrid juice. *Leaves.*—Compound or divided, toothed or cut. *Flowers.*—Yellow, clustered. *Calyx.*—Of two sepals falling early. *Corolla.*—Of four petals. *Stamens.*—Sixteen to twenty-four. *Pistil.*—One, with a two-lobed stigma. *Pod.*—Slender, linear.

The name of celandine must always suggest the poet who never seemed to weary of writing in its honor :

> Pansies, lilies, kingcups, daisies,
> Let them live upon their praises ;
> Long as there's a sun that sets,
> Primroses will have their glory ;
> Long as there are violets,
> They will have a place in story ;
> There's a flower that shall be mine,
> 'Tis the little celandine.

And when certain yellow flowers which frequent the village roadside are pointed out to us as those of the celandine, we feel a sense of disappointment that the favorite theme of Wordsworth should arouse within us so little enthusiasm. So perhaps we are rather relieved than otherwise to realize that the botanical name

PLATE XXXVI

Bulb.

YELLOW ADDER'S TONGUE.—*E. Americanum.*

117

of this plant signifies *greater* celandine ; for we remember that the poet never failed to specify the *small* celandine as the object of his praise. The small celandine is *Ranunculus ficaria*, one of the Crowfoot family, and is only found in this country as an escape from gardens.

Gray tells us that the generic name, *Chelidonium*, from the ancient Greek for swallow, was given " because its flowers appear with the swallows ; " but if we turn to Gerarde we read that the title was not bestowed " because it first springeth at the coming in of the swallowes, or dieth when they go away, for as we have saide, it may be founde all the yeare ; but because some holde opinion, that with this herbe the dams restore sight to their young ones, when their eies be put out."

CELANDINE POPPY.

Stylophorum diphyllum.   Poppy Family.

Stem.—Low, two-leaved. *Stem-leaves.*—Opposite, deeply incised. *Root-leaves.*—Incised or divided. *Flowers.*—Deep yellow, large, one or more at the summit of the stem. *Calyx.*—Of two hairy sepals. *Corolla.*—Of four petals. *Stamens.*—Many. *Pistil.*—One, with a two to four-lobed stigma.

In April or May, somewhat south and westward, the woods are brightened, and occasionally the hill-sides are painted yellow, by this handsome flower. In both flower and foliage the plant suggests the celandine.

DOWNY YELLOW VIOLET.

Viola pubescens.   Violet Family.

Stems.—Leafy above, erect. *Leaves.*—Broadly heart-shaped, toothed. *Flowers.*—Yellow, veined with purple, otherwise much like those of the common blue violet.

> When beechen buds begin to swell,
>   And woods the blue-bird's warble know,
> The yellow violet's modest bell
>   Peeps from the last year's leaves below,

sings Bryant, in his charming, but not strictly accurate poem, for the chances are that the " beechen buds " have almost burst into

PLATE XXXVII

DOWNY YELLOW VIOLET.— *V. pubescens.*

foliage, and that the "blue-bird's warble" has been heard for some time when these pretty flowers begin to dot the woods.

The lines which run :

> Yet slight thy form, and low thy seat,
> And earthward bent thy gentle eye,
> Unapt the passing view to meet,
> When loftier flowers are flaunting nigh,

would seem to apply more correctly to the round-leaved, *V. rotundifolia*, than to the downy violet, for although its large, flat shining leaves are somewhat conspicuous, its flowers are borne singly on a low scape, which would be less apt to attract notice than the tall, leafy flowering stems of the other.

COMMON CINQUEFOIL. FIVE FINGER.

Potentilla Canadensis. Rose Family.

Stem.—Slender, prostrate, or sometimes erect. *Leaves.*—Divided really into three leaflets, but apparently into five by the parting of the lateral leaflets. *Flowers.*—Yellow, growing singly from the axils of the leaves. *Calyx.*—Deeply five-cleft, with bracts between each tooth, thus appearing ten-cleft. *Corolla.*—Of five rounded petals. *Stamens.*—Many. *Pistils.*—Many in a head.

From spring to nearly midsummer the roads are bordered and the fields carpeted with the bright flowers of the common cinquefoil. The passer-by unconsciously betrays his recognition of some of the prominent features of the Rose family by often assuming that the plant is a yellow-flowered wild strawberry. Both of the English names refer to the pretty foliage, cinquefoil being derived from the French *cinque feuilles*. The generic name, *Potentilla*, has reference to the powerful medicinal properties formerly attributed to the genus.

SHRUBBY CINQUEFOIL. FIVE FINGER.

Potentilla fruticosa. Rose Family.

Stem.—Erect, shrubby, one to four feet high. *Leaves.*—Divided into five to seven narrow leaflets. *Flowers.*—Yellow, resembling those of the common cinquefoil.

Of all the cinquefoils perhaps this one most truly merits the title five finger. Certainly its slender leaflets are much more

PLATE XXXVIII

Leaf.

SHRUBBY CINQUEFOIL.—*P. fruticosa.*

finger-like than those of the common cinquefoil. It is not a common plant in most localities, but is very abundant among the Berkshire Hills.

SILVERY CINQUEFOIL.

Potentilla argentea. Rose Family.

Stems.—Ascending, branched at the summit, white, woolly. *Leaves.*— Divided into five wedge-oblong, deeply incised leaflets, which are green above, white with silvery wool, beneath.

The silvery cinquefoil has rather large yellow flowers which are found in dry fields throughout the summer as far south as New Jersey.

GOLDEN RAGWORT. SQUAW-WEED.

Senecio aureus. Composite Family (p. 13).

Stem.—One to three feet high. *Root-leaves.*—Rounded, the larger ones mostly heart-shaped, toothed, and long-stalked. *Stem-leaves.*—The lower lyre-shaped, the upper lance-shaped, incised, set close to the stem. *Flower-heads.*—Yellow, clustered, composed of both ray and disk-flowers.

A child would perhaps liken the flower of the golden ragwort to a yellow daisy. Stain yellow the white rays of the daisy, diminish the size of the whole head somewhat, and you have a pretty good likeness of the ragwort. There need be little difficulty in the identification of this plant—although there are several marked varieties—for its flowers are abundant in the early year, at which season but few members of the Composite family are abroad.

The generic name is from *senex*—an old man—alluding to the silky down of the seeds, which is supposed to suggest the silvery hairs of age.

Closely allied to the golden ragwort is the common groundsel, *S. vulgaris,* which is given as food to caged birds. The flower-heads of this species are without rays.

———   ———

Clintonia borealis. Lily Family.

Scape.—Five to eight inches high, sheathed at its base by the stalks of two to four large, oblong, conspicuous leaves. *Flowers.*—Greenish-yellow, rather large, rarely solitary. *Perianth.*—Of six sepals. *Stamens.*—Six, protruding. *Pistil.*—One, protruding. *Fruit.*—A blue berry.

When rambling through the cool, moist woods our attention is often attracted by patches of great dark, shining, leaves ; and

PLATE XXXIX

Fruit.

Clintonia borealis.

123

if it be late in the year we long to know the flower of which this rich foliage is the setting. To satisfy our curiosity we must return the following May or June, when we shall probably find that a slender scape rises from its midst bearing at its summit several bell-shaped flowers, which, without either high color or fragrance, are peculiarly charming. It is hard to understand why this beautiful plant has received no English name. As to its generic title we cannot but sympathize with Thoreau. "Gray should not have named it from the Governor of New York," he complains; "what is he to the lovers of flowers in Massachusetts? If named after a man, it must be a man of flowers. . . . Name your canals and railroads after Clinton, if you please, but his name is not associated with flowers."

C. umbellata is a more Southern species, with smaller white flowers, which are speckled with green or purplish dots.

YELLOW LADY'S SLIPPER. WHIP-POOR-WILL'S SHOE.
Cypripedium pubescens. Orchis Family (p. 17).

Stem.—About two feet high, downy, leafy to the top, one to three-flowered. *Leaves.*—Alternate, broadly oval, many-nerved and plaited. *Flowers.*—Large, yellow. *Perianth.*—Two of the three brownish, elongated sepals united into one under the lip; the lateral petals linear, wavy-twisted, brownish; the pale yellow lip an inflated pouch. *Stamens.*—Two, the short filaments of each bearing a two-celled anther. *Stigma.*—Broad, obscurely three-lobed, moist and roughish.

The yellow lady's slipper usually blossoms in May or June, a few days later than its pink sister, *C. acaule.* Regarding its favorite haunts, Mr. Baldwin * says: "Its preference is for maples, beeches, and particularly butternuts, and for sloping or hilly ground, and I always look with glad suspicion at a knoll covered with ferns, cohoshes, and trilliums, expecting to see a clump of this plant among them. Its sentinel-like habit of choosing 'sightly places' leads it to venture well up on mountain sides."

The long, wavy, brownish petals give the flower an alert, startled look when surprised in its lonely hiding-places.

C. parviflorum, the small yellow lady's slipper, differs from

* Orchids of New England.

PLATE XL

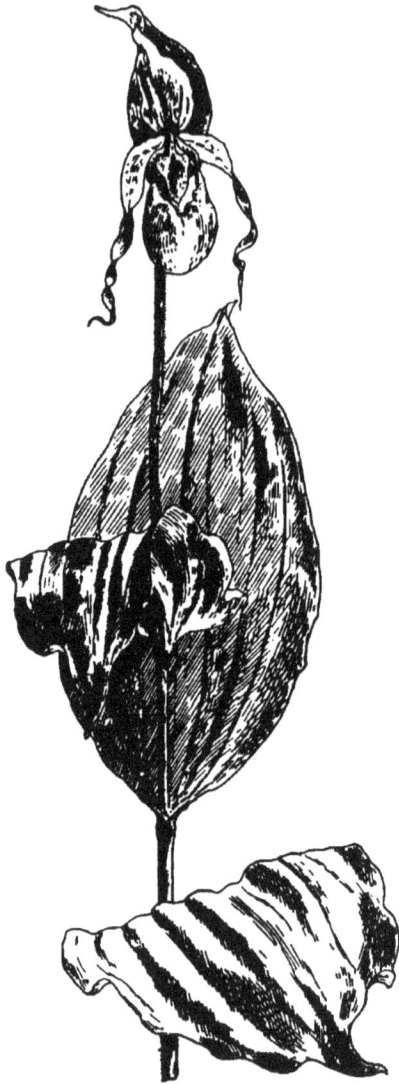

SMALLER YELLOW LADY'S SLIPPER.—*C. parviflorum.*

125

C. pubescens in the superior richness of its color as well as in its size. It also has the charm of fragrance.

EARLY MEADOW PARSNIP.

Zizia aurea. Parsley Family (p. 15).

One to three feet high. *Leaves.*—Twice or thrice-compound, leaflets oblong to lance-shaped, toothed. *Flowers.*—Yellow, small, in compound umbels.

This is one of the earliest members of the Parsley family to appear. Its golden flower-clusters brighten the damp meadows and the borders of streams in May or June and closely resemble the meadow parsnip, *Thaspium aureum*, of which this species was formerly considered a variety, of the later year.

The tall, stout, common wild parsnip, *Pastinaca sativa*, is another yellow representative of this family in which white flowers prevail, the three plants here mentioned being the only yellow species commonly encountered. The common parsnip may be identified by its grooved stem and simply compound leaves. Its roots have been utilized for food at least since the reign of Tiberius, for Pliny tells us that that Emperor brought them to Rome from the banks of the Rhine, where they were successfully cultivated.

GOLDEN CLUB.

Orontium aquaticum. Arum Family.

Scape.—Slender, elongated. *Leaves.*—Long-stalked, oblong, floating. *Flowers.*—Small, yellow, crowded over the narrow spike or spadix.

When we go to the bogs in May to hunt for the purple flower of the pitcher-plant we are likely to chance upon the well-named golden-club. This curious-looking club-shaped object, which is found along the borders of ponds, indicates its relationship to the jack-in-the-pulpit, and still more to the calla-lily, but unlike them its tiny flowers are shielded by no protecting spathe.

Kalm tells us in his "Travels," "that the Indians called the plant *Taw-Kee*, and used its dried seeds as food."

SPEARWORT.

Ranunculus ambigens. Crowfoot Family.

Stems.—One to two feet high. *Leaves.*—Oblong or lance-shaped, mostly toothed, contracted into a half-clasping leaf-stalk. *Flowers.*—Bright yellow, solitary or clustered. *Calyx.*—Of five sepals. *Corolla.*—Of five to seven oblong petals. *Stamens.*—Indefinite in number, occasionally few. *Pistils.*—Numerous in a head.

Many weeks after the marsh marigolds have passed away, just such marshy places as they affected are brightly flecked with gold. Wondering, perhaps, if they can be flowering for the second time in the season, we wade recklessly into the bog to rescue, not the marsh marigold, but its near relation, the spearwort, which is still more closely related to the buttercup, as a little comparison of the two flowers will show. This plant is especially common at the North.

INDIAN CUCUMBER-ROOT.

Medeola Virginica. Lily Family.

Root.—Tuberous, shaped somewhat like a cucumber, with a suggestion of its flavor. *Stem.*—Slender, from one to three feet high, at first clothed with wool. *Leaves.*—In two whorls on the flowering plants, the lower of five to nine oblong, pointed leaves set close to the stem, the upper usually of three or four much smaller ones. *Flowers.*—Greenish-yellow, small, clustered, recurved, set close to the upper leaves. *Perianth.*—Of three sepals and three petals, oblong and alike. *Stamens.*—Six, reddish-brown. *Pistil.*—With three stigmas, long, recurved, and reddish-brown. *Fruit.*—A purple berry.

One is more apt to pause in September to note the brilliant foliage and purple berries of this little plant than to gather the drooping inconspicuous blossoms for his bunch of wood-flowers in June. The generic name is after the sorceress Medea, on account of its supposed medicinal virtues, of which, however, there seems to be no record.

The tuberous rootstock has the flavor, and something the shape, of the cucumber, and was probably used as food by the Indians. It would not be an uninteresting study to discover which of our common wild plants are able to afford pleasant and nutritious food ; in such a pursuit many of the otherwise unattractive popular names would prove suggestive.

COMMON BLADDERWORT.

Utricularia vulgaris. Bladderwort Family.

Stems.—Immersed, one to three feet long. *Leaves.*—Many-parted, hair-like, bearing numerous bladders. *Scape.*—Six to twelve inches long. *Flowers.*—Yellow, five to twelve on each scape. *Calyx.*—Two-lipped. *Corolla.*—Two-lipped, spurred at the base. *Stamens.*—Two. *Pistil.*—One.

This curious water-plant may or may not have roots ; in either case it is not fastened to the ground, but is floated by means of the many bladders which are borne on its finely dissected leaves. It is commonly found in ponds and slow streams, flowering throughout the summer. Thoreau calls it " a dirty-conditioned flower, like a sluttish woman with a gaudy yellow bonnet."

The horned bladderwort, *U. cornuta*, roots in the peat-bogs and sandy swamps. Its large yellow helmet-shaped flowers are very fragrant, less than half a dozen being borne on each scape.

YELLOW POND-LILY. SPATTER DOCK.

Nuphar advena. Water-lily Family.

Leaves.—Floating or erect, roundish to oblong, with a deep cleft at their base. *Flowers.*—Yellow, sometimes purplish, large, somewhat globular. *Calyx.*—Of five or six sepals or more, yellow or green without. *Corolla.*—Of numerous small, thick, fleshy petals which are shorter than the stamens and resemble them. *Stamens.*—Very numerous. *Pistil.*—One, with a disk-like, many-rayed stigma.

Bordering the slow streams and stagnant ponds from May till August may be seen the yellow pond-lilies. These flowers lack the delicate beauty and fragrance of the white water-lilies ; having, indeed, either from their odor, or appearance, or the form of their fruit, won for themselves in England the unpoetic title of " brandy-bottle." Owing to their love of mud they have also been called " frog-lilies." The Indians used their roots for food.

PLATE XLI

Rootstock.   Fruit.

INDIAN CUCUMBER-ROOT.—*M. Virginiana.*

WINTER-CRESS, YELLOW ROCKET. HERB OF ST. BARBARA.

Barbarea vulgaris. Mustard Family (p. 17).

Stem.—Smooth. *Leaves.*—The lower lyre-shaped ; the upper ovate, toothed or deeply incised at their base. *Flowers.*—Yellow, growing in racemes. *Pod.*—Linear, erect or slightly spreading.

As early as May we find the bright flowers of the winter-cress along the roadside. This is probably the first of the yellow mustards to appear.

BLACK MUSTARD.

Brassica nigra. Mustard Family (p. 17).

Often several feet high. *Stem.*—Branching. *Leaves.*—The lower with a large terminal lobe and a few small lateral ones. *Flowers.*—Yellow, rather small, growing in a raceme. *Pods.*—Smooth, erect, appressed, about half an inch long.

Many are familiar with the appearance of this plant who are ignorant of its name. The pale yellow flowers spring from the waste places along the roadside and border the dry fields throughout the summer. The tall spreading branches recall the biblical description : "It groweth up, and becometh greater than all herbs, and shooteth out great branches ; so that the fowls of the air may lodge under the shadow of it."

This plant is extensively cultivated in Europe, its ground seeds forming the well-known condiment. The ancients used it for medicinal purposes. It has come across the water to us, and is a troublesome weed in many parts of the country.

WILD RADISH.

Raphanus Raphanistrum. Mustard Family (p. 17).

One to three feet high. *Leaves.*—Rough, lyre-shaped. *Flowers.*—Yellow, veiny, turning white or purplish ; larger than those of the black mustard, otherwise resembling them. *Pod.*—Often necklace-form by constriction between the seeds.

This plant is a troublesome weed in many of our fields. It is the stock from which the garden radish has been raised.

PLATE XLII

WINTER-CRESS.—*B. vulgaris.*

131

CYNTHIA. DWARF DANDELION.

Krigia Virginica. Composite Family (p. 13).

Stems.—Several, becoming branched, leafy. *Leaves.*—Earlier ones roundish ; the latter narrower and often cleft. *Flower-heads.*—Yellow, composed entirely of strap-shaped flowers. .

In some parts of the country these flowers are among the earliest to appear. They are found in New England, as well as south and westward.

The flowers of *K. amplexicaulis* appear later, and their range is a little farther south. Near Philadelphia great masses of the orange-colored blossoms and pale green stems and foliage line the railway embankments in June.

RATTLESNAKE-WEED.

Hieracium venosum. Composite Family (p. 13).

Stem or Scape.—One or two feet high, naked or with a single leaf, smooth, slender, forking above. *Leaves.*—From the root, oblong, often making a sort of flat rosette, usually conspicuously veined with purple. *Flower-heads.*—Yellow, composed entirely of strap-shaped flowers.

The loosely clustered yellow flower-heads of the rattlesnake-weed somewhat resemble small dandelions. They abound in the pine-woods and dry, waste places of early summer. The purple-veined leaves, whose curious markings give to the plant its common name, grow close to the ground and are supposed to be effi-cacious in rattlesnake bites. Here again crops out the old " doctrine of signatures," for undoubtedly this virtue has been attributed to the species solely on account of the fancied resem-blance between its leaves and the markings of the rattlesnake.

H. scabrum is another common species, which may be distin-guished from the rattlesnake-weed by its stout, leafy stem and un-veined leaves.

DANDELION.

Taraxacum officinale. Composite Family (p. 13).

If Emerson's definition of a weed, as a plant whose virtues have not yet been discovered, be correct, we can hardly place the dandelion in that category, for its young sprouts have been val-ued as a pot-herb, its fresh leaves enjoyed as a salad, and its dried roots used as a substitute for coffee in various countries and

PLATE XLIII

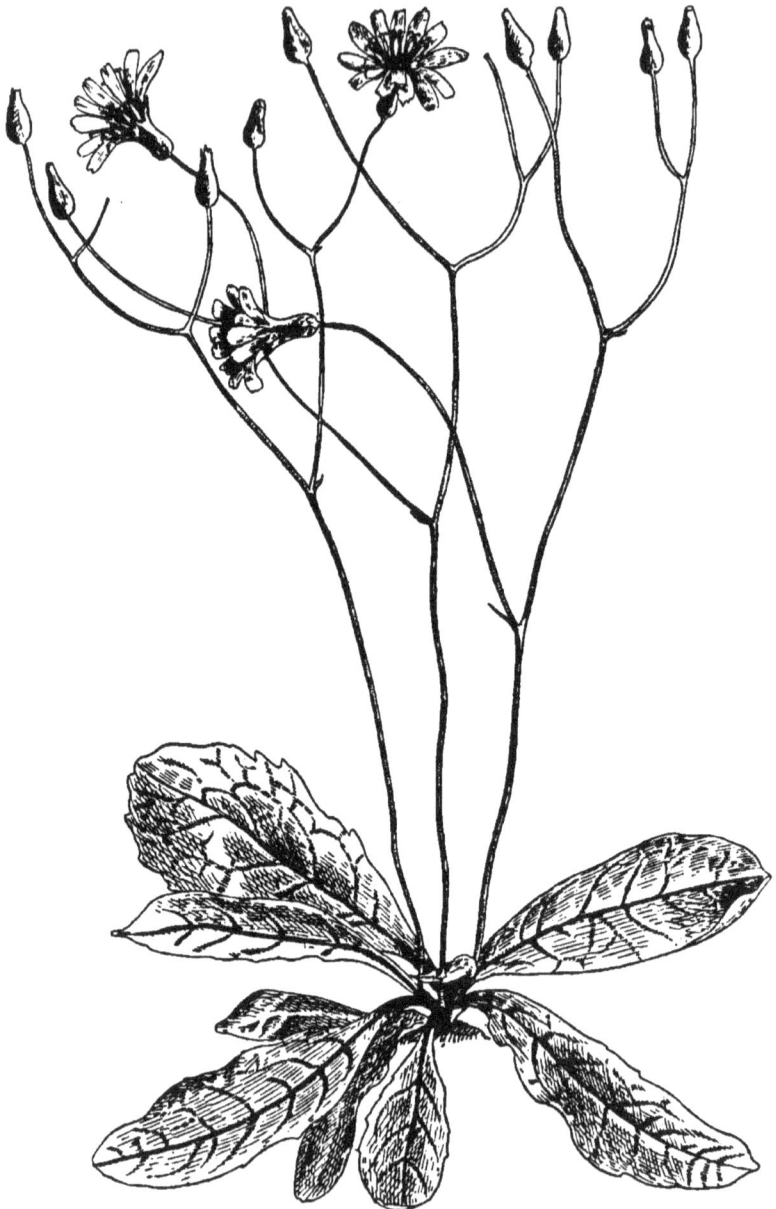

RATTLESNAKE-WEED.—*H. venosum.*

133

ages. It is said that the Apache Indians so greatly relish it as food, that they scour the country for many days in order to procure enough to appease their appetites, and that the quantity consumed by one individual exceeds belief. The feathery-tufted seeds which form the downy balls beloved as " clocks " by country children, are delicately and beautifully adapted to dissemination by the wind, which ingenious arrangement partly accounts for the plant's wide range. The common name is a corruption of the French *dent de lion*. There is a difference of opinion as to which part of the plant is supposed to resemble a lion's tooth. Some fancy the jagged leaves gave rise to the name, while others claim that it refers to the yellow flowers, which they liken to the golden teeth of the heraldic lion. In nearly every European country the plant bears a name of similar signification.

POVERTY-GRASS.

Hudsonia tomentosa. Rock-rose Family.

"Bushy, heath-like little shrubs, seldom a foot high." (Gray.) *Leaves.* —Small, oval or narrowly oblong, pressed close to the stem. *Flowers.*— Bright yellow, small, numerous, crowded along the upper part of the branches. *Calyx.*—Of five sepals, the two outer much smaller. *Corolla.*—Of five petals. *Stamens.*—Nine to thirty. *Pistil.*—One, with a long and slender style.

In early summer many of the sand-hills along the New England coast are bright with the yellow flowers of this hoary little shrub. It is also found as far south as Maryland and near the Great Lakes. Each blossom endures for a single day only. The plant's popular name is due to its economical habit of utilizing sandy unproductive soil where little else will flourish.

BUSH-HONEYSUCKLE.

Diervilla trifida. Honeysuckle Family.

An upright shrub from one to four feet high. *Leaves.*—Opposite, oblong, taper-pointed. *Flowers.*—Yellow, sometimes much tinged with red, clustered usually in threes, in the axils of the upper leaves and at the summit of the stem. *Calyx.*—With slender awl-shaped lobes. *Corolla.*—Funnel-form, five-lobed, the lower lobe larger than the others and of a deeper yellow, with a small nectar-bearing gland at its base. *Stamens.*—Five. *Pistil.*—One.

This pretty little shrub is found along our rocky hills and mountains. The blossoms appear in early summer, and form a

PLATE XLIV

BUSH-HONEYSUCKLE.—*D. trifida.*

good example of nectar-bearing flowers. The lower lobe of the corolla is crested and more deeply colored than the others, thus advising the bee of secreted treasure. The hairy filaments of the stamens are so placed as to protect the nectar from injury by rain. When the blossom has been despoiled and at the same time fertilized, for the nectar-seeking bee has probably deposited some pollen upon its pistil, the color of the corolla changes from a pale to a deep yellow, thus giving warning to the insect-world that further attentions would be useless to both parties.

COW WHEAT.

Melampyrum Americanum. Figwort Family.

Stem. — Low, erect, branching. *Leaves.* — Opposite, lance-shaped. *Flowers.*—Small, greenish-yellow, solitary in the axils of the upper leaves. *Calyx.*—Bell-shaped, four-cleft. *Corolla.*—Two-lipped, upper lip arched, lower three-lobed and spreading at the apex. *Stamens.*—Four. *Pistil.*— One.

In the open woods, from June until September, we encounter the pale yellow flowers of this rather insignificant little plant. The cow wheat was formerly cultivated by the Dutch as food for cattle. The Spanish name, *Trigo de Vaca*, would seem to indicate a similar custom in Spain. The generic name, *Melampyrum*, is from the Greek, and signifies *black wheat*, in reference to the appearance of the seeds of some species when mixed with grain. The flower would not be likely to attract one's attention were it not exceedingly common in some parts of the country, flourishing especially in our more eastern woodlands..

MEADOW LILY. WILD YELLOW LILY.

Lilium Canadense. Lily Family.

Stem.—Two to five feet high. *Leaves.*—Whorled, lance-shaped. *Flowers.*—Yellow, spotted with reddish-brown, bell-shaped, two to three inches long. *Perianth.*—Of six recurved sepals, with a nectar-bearing furrow at their base. *Stamens.*—Six, with anthers loaded with brown pollen. *Pistil.*—One, with a three-lobed stigma.

What does the summer bring which is more enchanting than a sequestered wood-bordered meadow hung with a thousand of these delicate, nodding bells which look as though ready to

PLATE XLV

MEADOW LILY.—*L. Canadense.*

137

tinkle at the least disturbance and sound an alarum among the flowers ?

These too are true "lilies of the field," less gorgeous, less imposing that the Turks' caps, but with an unsurpassed grace and charm of their own. "Fairy-caps," these pointed blossoms are sometimes called ; "witch-caps," would be more appropriate still. Indeed they would make dainty headgear for any of the dim inhabitants of Wonder-Land.

The growth of this plant is very striking when seen at its best. The erect stem is surrounded with regular whorls of leaves, from the upper one of which curves a circle of long-stemmed, nodding flowers. They suggest an exquisite design for a church candelabra.

PRICKLY PEAR. INDIAN FIG.

Opuntia Rafinesquii. Cactus Family.

Flowers.—Yellow, large, two and a half to three and a half inches across. *Calyx.*—Of numerous sepals. *Corolla.*—Of ten or twelve petals. *Stamens.* —Numerous. *Pistil.*—One, with numerous stigmas. *Fruit.*—Shaped like a small pear, often with prickles over its surface.

This curious looking plant is one of the·only two representatives of the Cactus family in the Northeastern States. It has deep green, fleshy, prickly, rounded joints and large yellow flowers, which are often conspicuous in summer in dry, sandy places along the coast.

O. vulgaris, the only other species found in Northeastern America, has somewhat smaller flowers, but otherwise so closely resembles *O. Rafinesquii* as to make it difficult to distinguish between the two.

FOUR-LEAVED LOOSESTRIFE.

Lysimachia quadrifolia. Primrose Family.

Stem.—Slender, one or two feet high. *Leaves.*—Narrowly oblong, whorled in fours, fives, or sixes. *Flowers.*—Yellow, spotted or streaked with red, on slender, hair-like flower-stalks from the axils of the leaves. *Calyx.*—Five or six-parted. *Corolla.*—Very deeply five or six-parted. *Stamens.*—Four or five. *Pistil.*—One.

This slender pretty plant grows along the roadsides and attracts one's notice in June by its regular whorls of leaves and

PLATE XLVI

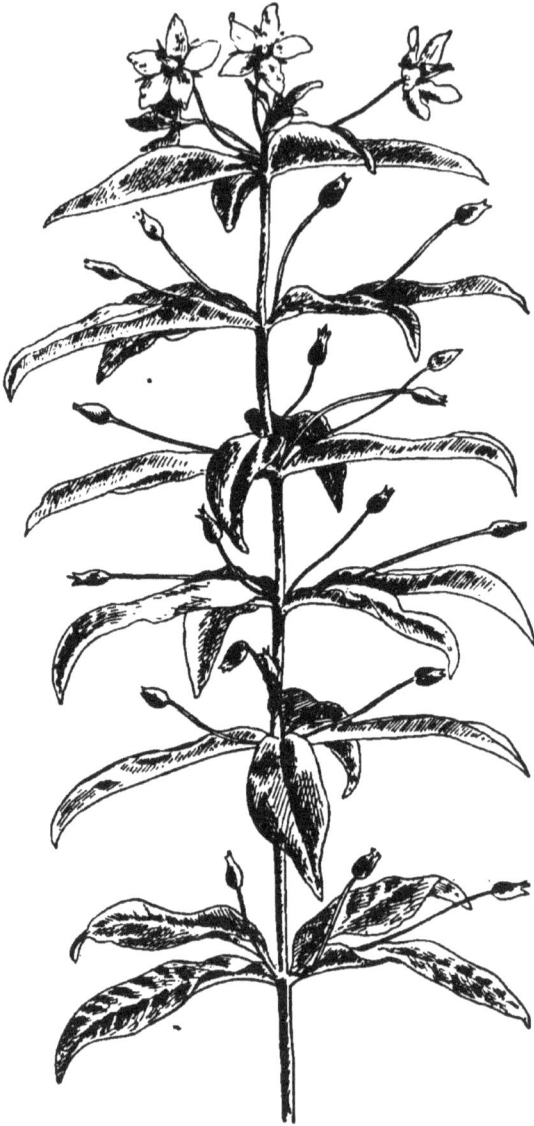

FOUR-LEAVED LOOSESTRIFE.—*L. quadrifolia.*

flowers. Linnæus says that this genus is named after Lysim-achus, King of Sicily. Loosestrife is the English for Lysim-achus; but whether the ancient superstition that the placing of these flowers upon the yokes of oxen rendered the beasts gentle and submissive arose from the peace-suggestive title or from other causes, I cannot discover.

YELLOW LOOSESTRIFE.

Lysimachia stricta. Primrose Family.

The yellow loosestrife bears its flowers, which are similar to those of *L. quadrifolia*, in a terminal raceme; it has opposite lance-shaped leaves. Its bright yellow clusters border the streams and brighten the marshes from June till August.

ROCK-ROSE. FROST-WEED.

Helianthemum Canadense. Rock-rose Family.

About one foot high. *Leaves.*—Set close to the stem, simple, lance-oblong. *Flowers.*—Of two kinds : the earlier, more noticeable ones, yellow, solitary, about one inch across ; the later ones small and clustered, usually without petals. *Calyx.*—(Of the petal-bearing flowers) of five sepals. *Corolla.*—Of five early falling petals which are crumpled in the bud. *Stamens.*—Numerous. *Pistil.*—One, with a three-lobed stigma.

These fragile bright yellow flowers are found in gravelly places in early summer. Under the influence of the sunshine they open once; by the next day their petals have fallen, and their brief beauty is a thing of the past. On June 17th Thoreau finds this " broad, cup-like flower, one of the most delicate yellow flowers, with large spring-yellow petals, and its stamens laid one way."

In the Vale of Sharon a nearly allied rose-colored species abounds. This is believed by some of the botanists who have travelled in that region to be the Rose of Sharon which Solomon has celebrated.

The name of frost-weed has been given to our plant because of the crystals of ice which shoot from the cracked bark at the base of the stem in late autumn.

PLATE XLVII

YELLOW LOOSESTRIFE.—*L. stricta.*

Steironema ciliatum. Primrose Family.

Stem.—Erect, two to four feet high. *Leaves.*—Opposite, narrowly oval, on fringed leaf-stalks. *Flowers.*—Yellow, on slender stalks from the axils of the leaves. *Calyx.*—Deeply five-parted. *Corolla.*—Deeply five-lobed, wheel-shaped, yellow, with a reddish centre. *Stamens.*—Five. *Pistil.*—One.

This plant is nearly akin to the yellow loosestrifes, but unfortunately it has no English name. It abounds in low grounds and thickets, putting forth its bright wheel-shaped blossoms early in July.

COMMON BARBERRY.

Berberis vulgaris. Barberry Family.

A shrub. *Leaves.*—Oblong, toothed, in clusters from the axil of a thorn. *Flowers.*—Yellow, in drooping racemes. *Calyx.*—Of six sepals, with from two to six bractlets without. *Corolla.*—Of six petals. *Stamens.*—Six. *Pistil.*—One. *Fruit.*—An oblong scarlet berry.

This European shrub has now become thoroughly wild and very plentiful in parts of New England. The drooping yellow flowers of May and June are less noticeable than the oblong clustered berries of September, which light up so many overgrown lanes, and often decorate our lawns and gardens as well.

The ancients extracted a yellow hair-dye from the barberry ; and to-day it is used to impart a yellow color to wool. Both its common and botanical names are of Arabic origin.

YELLOW STAR-GRASS.

Hypoxis erecta. Amaryllis Family.

Scapes.—Slender, few-flowered. *Leaves.*—Linear, grass-like, hairy. *Flowers.*—Yellow. *Perianth.*—Six-parted, spreading, the divisions hairy and greenish outside, yellow within. *Stamens.*—Six. *Pistil.*—One.

When our eyes fall upon what looks like a bit of evening sky set with golden stars, but which proves to be only a piece of shaded turf gleaming with these pretty flowers, we recall Longfellow's musical lines :

> Spake full well in language quaint and olden,
> One who dwelleth on the castled Rhine,
> When he called the flowers so blue and golden,
> Stars, which in earth's firmament do shine.

The plant grows abundantly in open woods and meadows, flowering in early summer.

PLATE XLVIII

YELLOW STAR-GRASS.—*H. erecta.*

143

WILD INDIGO.

Baptisia tinctoria. Pulse Family (p. 16).

Two or three feet high. *Stems.*—Smooth and slender. *Leaves.*—Divided into three rounded leaflets, somewhat pale with a whitish bloom, turning black in drying. *Flowers.*—Papilionaceous, yellow, clustered in many short, loose racemes. .

This rather bushy - looking, bright - flowered plant is constantly encountered in our rambles throughout the somewhat dry and sandy parts of the country in midsummer. It is said that it is found in nearly every State in the Union, and that it has been used as a homœopathic remedy for typhoid fever. Its young shoots are eaten at times in place of asparagus. Both the botanical and common names refer to its having yielded an economical but unsuccessful substitute for indigo.

YELLOW CLOVER. HOP CLOVER.

Trifolium agrarium. Pulse Family (p. 16).

Six to twelve inches high. *Leaves.*—Divided into three oblong leaflets. *Flowers.*—Papilionaceous, yellow, small, in close heads.

Although this little plant is found in such abundance along our New England roadsides and in many other parts of the country as well, comparatively few people seem to recognize it as a member of the clover group, despite a marked likeness in the leaves and blossoms to others of the same family.

The name clover probably originated in the Latin *clava*-clubs, in reference to the fancied resemblance between the three-pronged club of Hercules and the clover leaf. The clubs of our playing-cards and the *trèfle* (trefoil) of the French are probably an imitation of the same leaf.

The nonesuch, *Medicago lupulina*, with downy, procumbent stems, and flowers which grow in short spikes, is nearly allied to the hop clover. In its reputed superiority as fodder its English name is said to have originated. Dr. Prior says that for many years this plant has been recognized in Ireland as the true shamrock.

DYER'S GREEN-WEED. WOOD-WAXEN. NEW ENGLAND WHIN.

Genista tinctoria. Pulse Family (p. 16).

A shrubby plant from one to two feet high. *Leaves.*—Lance-shaped. *Flowers.*—Papilionaceous, yellow, growing in spiked racemes.

This is another foreigner which has established itself in Eastern New York and Massachusetts, where it covers the barren hill-sides with its yellow flowers in early summer. It is a common English plant, formerly valued for the yellow dye which it yielded. It is an undesirable intruder in pasture-lands, as it gives a bitter taste to the milk of cows which feed upon it.

YELLOW SWEET CLOVER. YELLOW MELILOT.

Melilotus officinalis. Pulse Family (p. 16).

Two to four feet high. *Stem.*—Upright. *Leaves.*—Divided into three, toothed leaflets. *Flowers.*—Papilionaceous, yellow, growing in spike-like racemes.

This plant is often found blossoming along the roadsides in early summer. It was formerly called in England "king's-clover," because, as Parkinson writes, "the yellowe flowers doe crown the top of the stalkes." The leaves become fragrant in drying.

RATTLEBOX.

Crotalaria sagittalis. Pulse Family (p. 16).

Stem.—Hairy, three to six inches high. *Leaves.*—Undivided, oval or lance-shaped. *Flowers.*—Papilionaceous, yellow, but few in a cluster. *Pod.*—Inflated, many-seeded, blackish.

The yellow flowers of the rattlebox are found in the sandy meadows and along the roadsides during the summer. Both the generic and English names refer to the rattling of the loose seeds within the inflated pod.

BUTTER-AND-EGGS. TOADFLAX.

Linaria vulgaris. Figwort Family.

Stem.—Smooth, erect, one to three feet high. *Leaves.*—Alternate, linear or nearly so. *Flowers.*—Of two shades of yellow, growing in terminal racemes. *Calyx.*—Five-parted. *Corolla.*—Pale yellow tipped with orange, long-spurred, two-lipped, closed in the throat. *Stamens.*—Four. *Pistil.*—One.

The bright blossoms of butter-and-eggs grow in full, close clusters which enliven the waste places along the roadside so commonly, that little attention is paid to these beautiful and conspicuous flowers. They would be considered a "pest" if they did not display great discrimination in their choice of locality, usually selecting otherwise useless pieces of ground. The common name of butter-and-eggs is unusually appropriate, for the two shades of yellow match perfectly their namesakes. Like nearly all our common weeds, this plant has been utilized in various ways by the country people. It yielded what was considered at one time a valuable skin lotion, while its juice mingled with milk constitutes a fly-poison. Its generic name, *Linaria*, and its English title, toadflax, arose from a fancied resemblance between its leaves and those of the flax.

WILD SENNA.

Cassia Marilandica. Pulse Family.

Stem.—Three or four feet high. *Leaves.*—Divided into from six to nine pairs of narrowly oblong leaflets. *Flowers.*—Yellow, in short clusters from the axils of the leaves. *Calyx.*—Of five sepals. *Corolla.*—Of five slightly unequal, spreading petals, usually somewhat spotted with reddish brown. *Stamens.*—Five to ten, unequal, some of them often imperfect. *Pistil.*—One. *Pod.*—Long and narrow, slightly curved, flat.

This tall, striking plant, with clusters of yellow flowers which appear in midsummer, grows abundantly along many of the New England roadsides, and also far south and west, thriving best in sandy soil. Although a member of the Pulse family its blossoms are not papilionaceous.

PLATE XLIX

BUTTER-AND-EGGS.—*L. vulgaris.*

PARTRIDGE-PEA.

Cassia Chamæcrista. Pulse Family.

Stems.—Spreading, eight inches to a foot long. *Leaves.*—Divided into from ten to fifteen pairs of narrow delicate leaflets, which close at night and are somewhat sensitive to the touch. *Flowers.*—Yellow, rather large and showy, on slender stalks beneath the spreading leaves ; not papilionaceous. *Calyx.*—Of five sepals. *Corolla.*—Of five rounded, spreading, somewhat unequal petals, two or three of which are usually spotted at the base with red or purple. *Stamens.*—Ten, unequal, dissimilar. *Pistil.*—One, with a slender style. *Pod.*—Flat.

The partridge-pea is closely related to the wild senna, and a pretty, delicate plant it is, with graceful foliage, and flowers in late summer which surprise us with their size, abounding in gravelly, sandy places where little else will flourish, brightening the railway embankments and the road's edge. It is at home all over the country south of Massachusetts and east of the Rocky Mountains, but it grows with a greater vigor and luxuriance in the south than elsewhere. The leaves can hardly be called sensitive to the touch, yet when a branch is snapped from the parent-stem or is much handled, the delicate leaflets will droop and fold, displaying their curious mechanism.

COMMON ST. JOHN'S-WORT.

Hypericum perforatum. St. John's-wort Family.

Stem.—Much branched. *Leaves.*—Small, opposite, somewhat oblong, with pellucid dots. *Flowers.*—Yellow, numerous, in leafy clusters. *Calyx.*—Of five sepals. *Corolla.*—Of five bright yellow petals, somewhat spotted with black. *Stamens.*—Indefinite in number. *Pistil.*—One, with three spreading styles.

"Too well known as a pernicious weed which it is difficult to extirpate," is the scornful notice which the botany gives to this plant whose bright yellow flowers are noticeable in waste fields and along roadsides nearly all summer. Its rank, rapid growth proves very exhausting to the soil, and every New England farmer wishes it had remained where it rightfully belongs— on the other side of the water.

Perhaps more superstitions have clustered about the St. John's-wort than about any other plant on record. It was formerly gathered on St. John's eve, and was hung at the doors and win-

PLATE L

COMMON ST. JOHN'S-WORT.—*H. perforatum.*

dows as a safeguard against thunder and evil spirits. A belief prevailed that on this night the soul had power to leave the body and visit the spot where it would be finally summoned from its earthly habitation, hence the all-night vigils which were observed at that time.

> The wonderful herb whose leaf will decide
> If the coming year shall make me a bride,

is the St. John's-wort, and the maiden's fate is favorably forecast by the healthy growth and successful blossoming of the plant which she has accepted as typical of her future.

In early times poets and physicians alike extolled its properties. An ointment was made of its blossoms, and one of its early names was " balm-of-the-warrior's-wound." It was considered so efficacious a remedy for melancholia that it was termed " fuga dæmonum." Very possibly this name gave rise to the general idea that it was powerful in dispelling evil spirits.

ST. ANDREW'S CROSS.

Ascyrum Crux-Andreæ. St. John's-wort Family.

Stem.—Low, branched. *Leaves.*— Opposite, narrowly oblong, black-dotted. *Flowers.*—Light-yellow. *Calyx.*—Of four sepals, the two outer broad and leaf-like, the inner much smaller. *Corolla.*—Of four narrowly oblong petals. *Stamens.*—Numerous. *Pistil.*—One, with two short styles.

From July till September these flowers may be found in the pine-barrens of New Jersey and farther south and westward, and on the island of Nantucket as well.

COMMON MULLEIN.

Verbascum Thapsus. Figwort Family.

Stem.—Tall and stout, from three to five feet high. *Leaves.*—Oblong, woolly. *Flowers.*—In a long dense spike. *Calyx.*—Five-parted. *Corolla.* —Yellow, with five slightly unequal rounded lobes. *Stamens.*—Ten, the three upper with white wool on their filaments. *Pistil.*—One.

The common mullein is a native of the island of Thapsos, from which it takes its specific name. It was probably brought to this country from Europe by the early colonists, notwithstanding the title of " American velvet plant," which it is rumored to bear in England. The Romans called it "candelaria," from

PLATE LI

COMMON MULLEIN.—*V. Thapsus.*

151

their custom of dipping the long dried stalk in suet and using it as a funeral torch, and the Greeks utilized the leaves for lamp-wicks. In more modern times they have served as a remedy for the pulmonary complaints of men and beasts alike, " mullein tea " being greatly esteemed by country people. Its especial efficacy with cattle has earned the plant its name of " bullocks' lungwort."

A low rosette of woolly leaves is all that can be seen of the mullein during its first year, the yellow blossoms on their long spikes opening sluggishly about the middle of the second sum-mer. It abounds throughout our dry, rolling meadows, and its tall spires are a familiar feature in the summer landscape.

MOTH MULLEIN.

Verbascum Blattaria. Figwort Family.

Stem.—Tall and slender. *Leaves.*—Oblong, toothed, the lower some-times lyre-shaped, the upper partly clasping. *Flowers.*—Yellow or white, tinged with red or purple, in a terminal raceme. *Calyx.*—Deeply five-parted. *Corolla.*—Butterfly-shape, of five rounded, somewhat unequal lobes. *Stamens.*—Five, with filaments bearded with violet wool and anthers loaded with orange-colored pollen. *Pistil.*—One.

Along the highway from July till October one encounters a slender weed on whose erect stem it would seem as though a number of canary-yellow or purplish-white moths had alighted for a moment's rest. These are the fragile, pretty flowers of the moth mullein, and they are worthy of a closer examination. The reddened or purplish centre of the corolla suggests the probabil-ity of hidden nectar, while the pretty tufts of violet wool borne by the stamens are well fitted to protect it from the rain. A little experience of the canny ways of these innocent-looking flowers lead one to ask the wherefore of every new feature.

YELLOW FRINGED ORCHIS. ORANGE ORCHIS.

Habenaria ciliaris. Orchis Family (p. 17).

Stem.—Leafy, one to two feet high. *Leaves.*—The lower oblong to lance-shaped, the upper passing into pointed bracts. *Flowers.*—Deep orange color, with a slender spur and deeply fringed lip ; growing in an ob-long spike.

Years may pass without our meeting this the most brilliant of our orchids. Suddenly one August day we will chance upon

PLATE LII

Single flower, enlarged.

YELLOW FRINGED ORCHIS.—*H. ciliaris.*

just such a boggy meadow as we have searched in vain a hundred times, and will behold myriads of its deep orange, dome-like spires erecting themselves in radiant beauty over whole acres of land. The separate flowers, with their long spurs and deeply fringed lips, will repay a close examination. They are well calculated, massed in such brilliant clusters, to arrest the attention of whatever insects may specially affect them. Although I have watched many of these plants I have never seen an insect visit one, and am inclined to think that they are fertilized by night moths.

Mr. Baldwin declares : "If I ever write a romance of Indian life, my dusky heroine, Birch Tree or Trembling Fawn, shall meet her lover with a wreath of this orchis on her head."

JEWEL-WEED. TOUCH-ME-NOT.
Geranium Family.
Impatiens pallida. Pale Touch-me-not.

Flowers.—Pale yellow, somewhat spotted with reddish-brown ; common northward.

Impatiens fulva. Spotted Touch-me-not.

Flowers.—Orange-yellow, spotted with reddish-brown ; common southward.

Two to six feet high. *Leaves.* — Alternate, coarsely toothed, oval. *Flowers.*—Nodding, loosely clustered, or growing from the axils of the leaves. *Calyx* and *Corolla.*—Colored alike, and difficult to distinguish ; of six pieces, the largest one extended backward into a deep sac ending in a little spur, the two innermost unequally two-lobed. *Stamens.*—Five, very short, united over the pistil. *Pistil.*—One.

These beautiful plants are found along shaded streams and marshes, and are profusely hung with brilliant jewel-like flowers during the summer months. In the later year they bear those closed inconspicuous blossoms which fertilize in the bud and are called cleistogamous flowers. The jewel-weed has begun to appear along the English rivers, and it is said that the ordinary showy blossoms are comparatively rare, while the cleistogamous ones abound. Does not this look almost like a determination on the part of the plant to secure a firm foothold in its new environment before expending its energy on flowers which, though radiant and attractive, are quite dependent on insect-visitors for fertilization and perpetuation ?

154

PLATE LIII

PALE JEWEL-WEED.—*I. pallida.*

The name touch-me-not refers to the seed-pods, which burst open with such violence when touched, as to project their seeds to a comparatively great distance. This ingenious mechanism secures the dispersion of the seeds without the aid of the wind or animals. In parts of New York the plant is called "silver-leaf," from its silvery appearance when touched with rain or dew, or when held beneath the water.

AGRIMONY.

Agrimonia Eupatoria. Rose Family.

One or two feet high. *Leaves.*—Divided into several coarsely toothed leaflets. *Flowers.*—Small, yellow, in slender spiked racemes. *Calyx.*—Five-cleft, beset with hooked teeth. *Corolla.*—Of five petals. *Stamens.*—Five to fifteen. *Pistils.*—One to four.

The slender yellow racemes of the agrimony skirt the woods throughout the later summer. In former times the plant was held in high esteem by town physician and country herbalist alike. Emerson longed to know

> Only the herbs and simples of the wood,
> Rue, cinquefoil, gill, vervain, and agrimony.

.Up to a recent date the plant has been dried and preserved by country people and might be seen exposed for sale in the shops of French villages. It has also been utilized in a dressing for shoe-leather. When about to flower it yields a pale yellow dye.

Chaucer calls it *egremoine.* The name is supposed to be derived from the Greek title for an eye-disease, for which the juice of a plant similarly entitled was considered efficacious. The crushed flower yields a lemon-like odor.

YELLOW WOOD SORREL.

Oxalis stricta. Geranium Family.

Stem.—Erect. *Leaves.*—Divided into three delicate clover-like leaflets. *Flowers.*—Golden-yellow. *Calyx.*—Of five sepals. *Corolla.*—Of five petals. *Stamens.*—Ten. *Pistil.*—One, with five styles.

All summer the small flowers of the yellow wood sorrel show brightly against their background of delicate leaves. The plant varies greatly in its height and manner of growth, flourishing abundantly along the roadsides. The small leaflets are open to the genial influence of sun and air during the hours of daylight, but at night they protect themselves from chill by folding one against another.

SUNDROPS.

Œnothera fruticosa. Evening Primrose Family.

This is a day-blooming species of the evening primrose, with large, pale yellow blossoms and alternate oblong or narrowly lance-shaped leaves, and of a much less rank habit. In early summer our roadsides are illuminated with these flowers.

Œnothera pumila is also a diurnal species. Its loosely spiked blossoms are much smaller than those of the sundrops.

EVENING PRIMROSE.

Œnothera biennis. Evening Primrose Family.

Stout, erect, one to five feet high. *Leaves.*—Alternate, lance-shaped to oblong. *Flowers.*—Pale yellow, in a leafy spike, opening at night. *Calyx.*—With a long tube, four-lobed. *Corolla.*—Of four somewhat heart-shaped petals. *Stamens.*—Eight, with long anthers. *Pistil.*—One, with a stigma divided into four linear lobes.

Along the roadsides in midsummer we notice a tall, rank-growing plant, which seems chiefly to bear buds and faded blossoms. And unless we are already familiar with the owl-like tendencies of the evening primrose, we are surprised, some dim twilight, to find this same plant resplendent with a mass of fragile yellow flowers, which are exhaling their faint delicious fragrance on the evening air.

One brief summer night exhausts the vitality of these delicate blossoms. The faded petals of the following day might serve as a text for a homily against all-night dissipation, did we not know that by its strange habit the evening primrose guards against the depredations of those myriad insects abroad during the day, which are unfitted to transmit its pollen to the pistil of another flower.

We are impressed by the utilitarianism in vogue in this floral world, as we note that the pale yellow of these blossoms gleams so vividly through the darkness as to advertise effectively their whereabouts, while their fragrance serves as a mute invitation to the pink night-moth, which is their visitor and benefactor. Why they change their habits in the late year and remain open during the day, I have not been able to discover.

HORSE BALM. RICH-WEED. STONE-ROOT.

Collinsonia Canadensis. Mint Family (p. 16).

One to three feet high. *Leaves.*—Opposite, large, ovate, toothed, pointed. *Flowers.*—Yellowish, lemon-scented, clustered loosely. *Calyx.*— Two-lipped, the upper lip three-toothed, the lower two-cleft. *Corolla.*— Elongated, somewhat two-lipped, the four upper lobes nearly equal, the lower large and long, toothed or fringed. *Stamens.*—Two (sometimes four, the upper pair shorter), protruding, diverging. *Pistil.*—One, with a two-lobed style.

In the damp rich woods of midsummer these strong-scented herbs, with their loose terminal clusters of lemon-colored, lemon-scented flowers are abundant. The plant was introduced into England by the amateur botanist and flower-lover, Collinson, after whom the species is named. The Indians formerly employed it as an application to wounds.

BLACK-EYED SUSAN. CONE-FLOWER.

Rudbeckia hirta. Composite Family (p. 13).

Stem.—Stout and hairy, one to two feet high. *Leaves.*—Rough and hairy, the upper long, narrow, set close to the stem ; the lower broader, with leaf-stalks. *Flower-heads.*—Composed of both ray and disk-flowers ; the former yellow, the latter brown and arranged on a cone-like receptacle.

By the middle of July our dry meadows are merry with black-eyed Susans, which are laughing from every corner and keeping up a gay midsummer carnival in company with the yellow lilies and brilliant milkweeds. They seem to revel in the long days of blazing sunlight, and are veritable salamanders among the flowers. Although now so common in our eastern fields they were first brought to us with clover-seed from the

PLATE LIV

EVENING PRIMROSE.—*Œ. biennis.*

West, and are not altogether acceptable guests, as they bid fair to add another anxiety to the already harassed life of the New England farmer.

———— ————

Rudbeckia laciniata. Composite Family (p. 13).

Two to seven feet high. *Stem.*—Smooth, branching. *Leaves.*—The lower divided into lobed leaflets, the upper irregularly three to five-parted. *Flower-heads.*—Yellow, rather large, composed of both ray and disk-flowers, the former drooping and yellow, the later dull greenish and arranged on a columnar receptacle.

This graceful, showy flower is even more decorative than the black-eyed Susan. Its drooping yellow rays are from one to two inches long. It may be found throughout the summer in the low thickets which border the swamps and meadows.

GOLDEN ASTER.

Chrysopsis Mariana. Composite Family (p. 13).

Stem.—Silky, with long weak hairs when young. *Leaves.*—Alternate, oblong. *Flower-heads.*—Golden-yellow, rather large, composed of both ray and disk-flowers.

In dry places along the roadsides of Southern New York and farther south, one can hardly fail to notice in late summer and autumn the bright clusters of the golden aster.

C. falcata is a species which may be found in dry sandy soil as far north as Massachusetts, with very woolly stems, crowded linear leaves, and small, clustered flower-heads.

GOLDEN-ROD.

Solidago. Composite Family (p. 13).

Flower-heads.—Golden-yellow, composed of both ray and disk-flowers.

About eighty species of golden-rod are native to the United States: of these forty-two species can be found in our North-eastern States. Many of them are difficult of identification, and it would be useless to describe any but a few of the more conspicuous forms.

PLATE LV

BLACK-EYED SUSAN.—*R. hirta.*

A common and noticeable species which flowers early in August is *S. Canadensis*, with a tall stout stem from three to six feet high, lance-shaped leaves, which are usually sharply toothed and pointed, and small flower-heads clustered along the branches which spread from the upper part of the stem.

Another early flowering species is *S. rugosa*. This is a lower plant than *S. Canadensis*, with broader leaves. Still another is the dusty golden-rod, *S. nemoralis*, which has a hoary aspect and very bright yellow flowers which are common in dry fields.

S. lanceolata has lance-shaped or linear leaves, and flowers which grow in flat-topped clusters, unlike other members of the family; the information that this is a golden-rod often creates surprise, as for some strange reason it seems to be confused with the tansy.

The sweet golden-rod, *S. odorata*, is easily recognized by its fragrant, shining, dotted leaves. *S. cæsia*, or the blue-stemmed, is a wood-species and among the latest of the year, putting forth its bright clusters for nearly the whole length of its stem long after many of its brethren look like brown wraiths of their former selves. The silver rod, *S. bicolor*, whose whitish flowers are a departure from the family habit, also survives the early cold and holds its own in the dry woods.

The only species native to Great Britain is *S. Virga-aurea*.

The generic name is from two Greek words which signify *to make whole*, and refer to the healing properties which have been attributed to the genus.

ELECAMPANE.

Inula Helenium. Composite Family (p. 13).

Stem.—Stout, three to five feet high. *Leaves.*—Alternate, large, woolly beneath, the upper partly clasping. *Flower-heads.*—Yellow, large, composed of both ray and disk-flowers.

When we see these great yellow disks peeping over the pasture walls or flanking the country lanes, we feel that midsummer is at its height. Flowers are often subservient courtiers, and make acknowledgment of whatever debt they owe by that subtlest of flatteries—imitation. Did not the blossoms of the

PLATE LVI

ELECAMPANE.—*I. Helenium.*

163

dawning year frequently wear the livery of the snow which had thrown its protecting mantle over their first efforts? And these newcomers—whose gross, rotund countenances so clearly betray the results of high living—do not they pay their respects to their great benefactor after the same fashion?—with the result that a myriad miniature suns shine upward from meadow and roadside.

The stout, mucilaginous root of this plant is valued by farmers as a horse-medicine, especially in epidemics of epizootic, one of its common names in England being horse-heal.

In ancient times the elecampane was considered an important stimulant to the human brain and stomach, and it was mentioned as such in the writings of Hippocrates, the "Father of Medicine," over two thousand years ago.

The common name is supposed to be a corruption of *ala campania*, and refers to the frequent occurrence of the plant in that ancient province of Southern Italy.

FALL DANDELION.

Leontodon autumnalis. Composite Family (p. 13).

Scape.—Five to fifteen inches high, branching. *Leaves.*—From the root, toothed or deeply incised. *Flower-heads.*—Yellow, composed entirely of strap-shaped flowers ; smaller than those of the common dandelion.

From June till November we find the fall dandelion along the New England roadsides, as well as farther south. While the yellow flower-heads somewhat suggest small dandelions the general habit of the plant recalls some of the hawkweeds.

WILD SUNFLOWER.

Helianthus giganteus. Composite Family (p. 13).

Stem.—Rough or hairy, from three to ten feet high, branched above. *Leaves.*—Lance-shaped, pointed, rough to the touch, set close to the stem. *Flower-heads.*—Yellow, composed of both ray and disk-flowers.

In late summer many of our lanes are hedged by this beautiful plant, which, like other members of its family, lifts its yellow flowers sunward in pale imitation of the great lifegiver itself. We have twenty-two different species of sunflower. *H. divari-*

PLATE LVII

WILD SUNFLOWER.—*H. giganteus.*

catus is of a lower growth, with opposite, widely spreading leaves and larger flower-heads. *H. annuus* is the garden species familiar to all; this is said to be a native of Peru. Mr. Ellwanger writes regarding it : " In the mythology of the ancient Peruvians it occupied an important place, and was employed as a mystic decoration in ancient Mexican sculpture. Like the lotus of the East, it is equally a sacred and an artistic emblem, figuring in the symbolism of Mexico and Peru, where the Spaniards found it rearing its aspiring stalk in the fields, and serving in the temples as a sign and a decoration, the sun-god's officiating handmaidens wearing upon their breasts representations of the sacred flower in beaten gold."

Gerarde describes it as follows : " The Indian Sun or the golden floure of Peru is a plant of such stature and talnesse that in one Sommer, being sowne of a seede in April, it hath risen up to the height of fourteen foot in my garden, where one floure was in weight three pound and two ounces, and crosse overthwart the floure by measure sixteen inches broad."

The generic name is from *helios*—the sun, and *anthos*—a flower.

SNEEZEWEED. SWAMP SUNFLOWER.

Helenium autumnale. Composite Family (p. 13).

One to six feet high. *Stem.*—Angled, erect, branching. *Leaves.*—Alternate, lance-shaped. *Flower-heads.*—Yellow, composed of both ray and disk-flowers, the rays being somewhat cleft.

As far north as Connecticut we see masses of these bright flowers bordering the streams and swamps in September.

STICK-TIGHT. BUR MARIGOLD, BEGGAR-TICKS.

Bidens frondosa. Composite Family (p. 13).

Two to six feet high. *Stem.*—Branching. *Leaves.*—Opposite, three to five-divided. *Flower-heads.*—Consisting of brownish-yellow tubular flowers, with a leaf-like involucre beneath.

If one were only describing the attractive wild flowers, the stick-tight would certainly be omitted, as its appearance is not prepossessing, and the small barbed seed-vessels so cleverly fulfil

PLATE LVIII

Barbed fruit.

STICK-TIGHT.—*B. frondosa.*

their destiny in making one's clothes a means of conveyance to "fresh woods and pastures new" as to cause all wayfarers heartily to detest them. "How surely the desmodium growing on some cliff-side, or the bidens on the edge of a pool, prophesy the coming of the traveller, brute or human, that will transport their seeds on his coat," writes Thoreau. But the plant is so constantly encountered in late summer, and yet so generally unknown, that it can hardly be overlooked.

The larger bur marigold, *B. chrysanthemoides,* does its best to retrieve the family reputation for ugliness, and surrounds its dingy disk-flowers with a circle of showy golden rays which are strictly decorative, having neither pistils nor stamens, and leaving all the work of the household to the less attractive but more useful disk-flowers. Their effect is pleasing, and late into the autumn the moist ditches look as if sown with gold through their agency. The plant varies in height from six inches to two feet. Its leaves are opposite, lance-shaped, and regularly toothed.

SMOOTH FALSE FOXGLOVE.

Gerardia quercifolia. Figwort Family.

Stem.—Smooth, three to six feet high, usually branching. *Leaves.*—The lower usually deeply incised, the upper narrowly oblong, incised, or entire. *Flowers.*—Yellow, large, in a raceme or spike. *Calyx.*—Five-cleft. *Corolla.*—Two inches long, somewhat tubular, swelling above, with five more or less unequal, spreading lobes, woolly within. *Stamens.*—Four, in pairs, woolly. *Pistil.*—One.

These large pale yellow flowers are very beautiful and striking when seen in the dry woods of late summer. They are all the more appreciated because there are few flowers abroad at this season save the Composites, which are decorative and radiant enough, but usually somewhat lacking in the delicate charm we look for in a flower.

The members of this genus, which is named after Gerarde, the author of the famous "Herball," are supposed to be more or less parasitic in their habits, drawing their nourishment from the roots of other plants.

The downy false foxglove, *G. flava,* is usually a somewhat

168

PLATE LIX

SMOOTH FALSE FOXGLOVE.—*G. quercifolia.*

169

lower plant, with a close down, a less-branched stem, more entire leaves, and smaller, similar flowers.

TANSY.

Tanacetum vulgare. Composite Family (p. 13).

Stem.—Two to four feet high. *Leaves.*—Divided into toothed leaflets. *Flower-heads.*—Yellow, composed of tiny flowers which are nearly, if not all, tubular in shape ; borne in flat-topped clusters.

With the name of tansy we seem to catch a whiff of its strong-scented breath and a glimpse of some New England homestead beyond whose borders it has strayed to deck the roadside with its deep yellow, flat-topped flower-clusters. The plant has been used in medicine since the Middle Ages, and in more recent times it has been gathered by the country people for "tansy wine" and "tansy tea." In the Roman Church it typifies the bitter herbs which were to be eaten at the Paschal season ; and cakes made of eggs and its leaves are called "tansies," and eaten during Lent. It is also frequently utilized in more secular concoctions.

The common name is supposed to be a corruption of the Greek word for *immortality*.

WITCH-HAZEL.

Hamamelis Virginiana. Witch-hazel Family.

A tall shrub. *Leaves.*—Oval, wavy-toothed, mostly falling before the flowers appear. *Flowers.*—Honey-yellow, clustered, autumnal. *Calyx.*—Four-parted. *Corolla.*—Of four long narrow petals. *Stamens.*—Eight. *Pistils.*—Two. *Fruit.*—A capsule which bursts elastically, discharging its large seeds with vigor.

It seems as though the flowers of the witch-hazel were fairly entitled to the "booby-prize." of the vegetable world. Surely no other blossoms make their first appearance so invariably late upon the scene of action. The fringed gentian often begins to open its "meek and quiet eye" quite early in September. Certain species of golden-rod and aster continue to flower till late in the year, but they began putting forth their bright clusters before the summer was fairly over ; while the elusively fra-

PLATE LX

TANSY.—*T. vulgare.*

171

grant, pale yellow blossoms of the witch-hazel need hardly be expected till well on in September, when its leaves have fluttered earthward and its fruit has ripened. Does the pleasure which we experience at the spring-like apparition of this leafless yellow-flowered shrub in the autumn woods arise from the same depraved taste which is gratified by strawberries at Christmas, I wonder? Or is it that in the midst of death we have a foretaste of life; a prophecy of the great yearly resurrection which even now we may anticipate?

Thoreau's tastes in such directions were certainly not depraved, and he writes: "The witch-hazel loves a hill-side with or without woods or shrubs. It is always pleasant to come upon it unexpectedly as you are threading the woods in such places. Methinks I attribute to it some elfish quality apart from its fame. I love to behold its gray speckled stems." Under another date he writes: "Heard in the night a snapping sound, and the fall of some small body on the floor from time to time. In the morning I found it was produced by the witch-hazel nuts on my desk springing open and casting their seeds quite across my chamber, hard and stony as these nuts were."

The Indians long ago discovered the value of its bark for medicinal purposes, and it is now utilized in many well-known extracts. The forked branches formerly served as divining-rods in the search for water and precious ores. This belief in its mysterious power very possibly arose from its suggestive title, which Dr. Prior says should be spelled *wych*-hazel, as it was called after the wych-elm, whose leaves it resembles, and which was so named because the chests termed in old times "wyches" were made of its wood—

> His hall rofe was full of bacon flytches,
> The chambre charged was with wyches
> Full of egges, butter, and chese.*

NOTE.—The flowers of the American Woodbine and of the Fly Honeysuckle (p. 228), and of the Golden Corydalis (p. 192) are also yellow.

* Hazlitt's Early Popular Poetry.

III

PINK

TRAILING ARBUTUS. MAYFLOWER. GROUND LAUREL.

Epigæa repens. Heath Family.

Stem.—With rusty hairs, prostrate or trailing. *Leaves.*—Rounded, heart-shaped at base, evergreen. *Flowers.*—Pink, clustered, fragrant. *Calyx.*—Of five sepals. *Corolla.*—Five-lobed, salver-shaped, with a slender tube which is hairy within. *Stamens.*—Ten. *Pistil.*—One, with a five-lobed stigma.

> Pink, small, and punctual,
> Aromatic, low,

describes, but does scant justice to the trailing arbutus, whose waxy blossoms and delicious breath are among the earliest prophecies of perfume-laden summer. We look for these flowers in April—not beneath the snow—where tradition rashly locates them—but under the dead brown leaves of last year; and especially among the pines and in light sandy soil. Appearing as they do when we are eager for some tangible assurance that

> —the Spring comes slowly up this way,

they win from many of us the gladdest recognition of the year.

In New England they are called Mayflowers, being peddled about the streets of Boston every spring, under the suggestive and loudly emphasized title of " Ply-y-mouth Ma-ayflowers ! " Whether they owe this name to the ship which is responsible for so much, or to their season of blooming, in certain localities, might remain an open question had we not the authority of Whittier for attributing it to both causes. In a note prefacing " The Mayflowers," the poet says : " The trailing arbutus or Mayflower grows abundantly in the vicinity of Plymouth, and

173

was the first flower to greet the Pilgrims after their fearful winter.'' In the poem itself he wonders what the old ship had

> Within her ice-rimmed bay
> In common with the wild-wood flowers,
> The first sweet smiles of May?

and continues—

> Yet " God be praised ! " the Pilgrim said,
> Who saw the blossoms peer
> Above the brown leaves, dry and dead,
> " Behold our Mayflower here ! "
>
> God wills it, here our rest shall be,
> Our years of wandering o'er,
> For us the Mayflower of the sea
> Shall spread her sails no more.
>
> O sacred flowers of faith and hope,
> As sweetly now as then,
> Ye bloom on many a birchen slope,
> In many a pine-dark glen.
> .   .   .   .   .   .
> So live the fathers in their sons,
> Their sturdy faith be ours,
> And ours the love that overruns
> Its rocky strength with flowers.

If the poet's fancy was founded on fact, and if our lovely and widespread Mayflower was indeed the first blossom noted and christened by our forefathers, it seems as though the problem of a national flower must be solved by one so lovely and historic as to silence all dispute. And when we read the following prophetic stanzas which close the poem, showing that during another dark period in·our nation's history these brave little blossoms, struggling through the withered leaves, brought a message of hope and courage to the heroic heart of the Quaker poet, our feeling that they are peculiarly identified with our country's perilous moments is intensified :

> The Pilgrim's wild and wintry day
> At shadow round us draws ;
> The Mayflower of his stormy bay
> Our Freedom's struggling cause.
>
> But warmer suns erelong shall bring
> To life the frozen sod ;
> And, through dead leaves of hope shall spring
> Afresh the flowers of God !

PLATE LXI

TRAILING ARBUTUS.—*E. repens.*

TWIN-FLOWER.—*L. borealis.*

TWIN–FLOWER.

Linnæa borealis. Honeysuckle Family.

Stem.—Slender, creeping and trailing. *Leaves.*—Rounded, shining and evergreen. *Flowers.*—Growing in pairs, delicate pink, fragrant, nodding on thread-like, upright flower-stalks. *Calyx.*—Five-toothed. *Corolla.*—Narrowly bell-shaped, five-lobed, hairy within. *Stamens.*—Four, two shorter than the others. *Pistil.*—One.

Whoever has seen

> —beneath dim aisles, in odorous beds,
> The slight Linnæa hang its twin-born heads,*

will not soon forget the exquisite carpeting made by its nodding pink flowers and dark shining leaves ; or the delicious perfume which actually filled the air and drew one's attention to the spot from which it was exhaled, tempting one to exclaim with Richard Jefferies, "Sweetest of all things is wild-flower air ! " That this little plant should have been selected as " the monument of the man of flowers " by the great Linnæus himself, bears testimony to his possession of that appreciation of the beautiful which is supposed to be lacking in men of long scientific training. I believe that there is extant at least one contemporary portrait of Linnæus in which he wears the tiny flowers in his buttonhole. The rosy twin-blossoms are borne on thread-like, forking flower-stalks, and appear in June in the deep, cool, mossy woods of the North.

SHOWY ORCHIS.

Orchis spectabilis. Orchis Family (p. 17).

Stem.—Four-angled, with leaf-like bracts, rising from fleshy, fibrous roots. *Leaves.*—Two, oblong, shining, three to six inches long. *Flowers.*—In a loose spike, purple-pink, the lower lip white.

This flower not only charms us with its beauty when its clusters begin to dot the rich May woods, but interests us as being usually the first member of the Orchis family to appear upon the scene ; although it is claimed in certain localities that the beautiful Calypso always, and the Indian moccason occasionally, precedes it.

* Emerson.

PLATE LXII

SHOWY ORCHIS.—*O. spectabilis.*

A certain fascination attends the very name of orchid. Botanist and unscientific flower-lover alike pause with unwonted interest when the discovery of one is announced. With the former there is always the possibility of finding some rare species, while the excitement of the latter is apt to be whetted with the hope of beholding a marvellous imitation of bee or butterfly fluttering from a mossy branch with roots that draw their nourishment from the air! While this little plant is sure to fail of satisfying the hopes of either, it is far prettier if less rare than many of its brethren, and its interesting mechanism will repay our patient study. It is said closely to resemble the "long purples," *O. mascula*, which grew near the scene of Ophelia's tragic death.

TWISTED STALK.

Streptopus roseus. Lily Family.

Stems.—Rather stout and zigzag, forking and diverging. *Leaves.*— Taper-pointed, slightly clasping. *Flowers.*—Dull purplish-pink, hanging on thread-like flower-stalks from the axils of the leaves. *Perianth.*—Somewhat bell-shaped, of six distinct sepals. *Stamens.*—Six. *Pistil.*—One, with a three-cleft stigma.

This plant presents a graceful group of forking branches and pointed leaves. No blossom is seen from above, but on picking a branch one finds beneath each of its outspread leaves one or two slender, bent stalks from which hang the pink, bell-like flowers. In general aspect the plant somewhat resembles its relations, the Solomon's seal, with which it is found blossoming in the woods of May or June. The English title is a translation of the generic name, *Streptopus.*

WILD PINK.

Silene Pennsylvanica. Pink Family.

Stems.—Four to eight inches high. *Leaves.*—Those from the root narrowly wedge-shaped, those on the stem lance-shaped, opposite. *Flowers.*— Bright pink, clustered. *Calyx.*—Five-toothed. *Corolla.*—Of five petals. *Stamens.*—Ten. *Pistil.*—One, with three styles.

When a vivid cluster of wild pinks gleams from some rocky opening in the May woods, it is difficult to restrain one's eager-

178

•

PLATE LXIII

Fruit.

TWISTED STALK.—*S. roseus.*

179

ness, for there is something peculiarly enticing in these fresh, vigorous-looking flowers. They are quite unlike most of their fragile contemporaries, for they seem to be already imbued with the glowing warmth of summer, and to have no memory of that snowy past which appears to leave its imprint on so many blossoms of the early year.

In waste places, from June until September or later, we find the small clustered pink flowers, which open transiently in the sunshine of the sleepy catchfly, *S. antirrhina.*

PINK LADY'S SLIPPER.   MOCCASON–FLOWER.

Cypripedium acaule.   Orchis Family (p. 17).

Scape.—Eight to twelve inches high, two-leaved at base, downy, one-flowered. *Leaves.*—Two, large, many-nerved and plaited, sheathing at the base. *Flower.*—Solitary, purple-pink. *Perianth.*—Of three greenish spreading sepals, the two lateral petals narrow, spreading, greenish, the pink lip in the shape of a large inflated pouch. *Stamens.*—Two, the short filaments each bearing a two-celled anther. *Stigma.*—Broad, obscurely three-lobed, moist and roughish.

> Graceful and tall the slender, drooping stem,
>   With two broad leaves below,
> Shapely the flower so lightly poised between,
>   And warm her rosy glow,

writes Elaine Goodale of the moccason-flower. This is a blossom whose charm never wanes. It seems to be touched with the spirit of the deep woods, and there is a certain fitness in its Indian name, for it looks as though it came direct from the home of the red man. All who have found it in its secluded haunts will sympathize with Mr. Higginson's feeling that each specimen is a rarity, even though he should find a hundred to an acre. Gray assigns it to "dry or moist woods," while Mr. Baldwin writes : "The finest specimens I ever saw sprang out of cushions of crisp reindeer moss high up among the rocks of an exposed hill-side, and again I have found it growing vigorously in almost open swamps, but nearly colorless from excessive moisture." The same writer quotes a lady who is familiar with it in the Adirondacks. She says : "It seems to have a great fondness for decaying wood, and I often see a whole row perched like

PLATE LXIV

PINK LADY'S SLIPPER.—*C. acaule.*

181

birds along a crumbling log." While I recall a mountain lake where the steep cliffs rise from the water's edge, here and there, on a tiny shelf strewn with pine-needles, can be seen a pair of large veiny leaves, above which, in early June, the pink balloon-like blossom floats from its slender scape.

--- ---

Calopogon pulchellus. Orchis Family (p. 17).

Scape.—Rising about one foot from a small solid bulb. *Leaf.*—Linear, grass-like. *Flowers.*—Two to six on each scape, purple-pink, about one inch broad, the lip as if hinged at its insertion, bearded toward the summit with white, yellow, and purple hairs. The peculiarity of this orchid is that the ovary is not twisted, and consequently the lip is on the upper instead of the lower side of the flower.

One may hope to find these bright flowers growing side by side with the glistening sundew in the rich bogs of early summer. Mr. Baldwin assigns still another constant companion to the *Calopogon*, an orchid which staggers under the terrifying title of *Pogonia ophioglossoides.* The generic name of *Calopogon* is from two Greek words signifying *beautiful beard* and has reference to the delicately bearded lip.

PINK AZALEA. WILD HONEYSUCKLE. PINXTER FLOWER. SWAMP PINK.

Rhododendron nudiflorum. Heath Family.

A shrub from two to six feet high. *Leaves.*—Narrowly oblong, downy underneath, usually appearing somewhat later than the flowers. *Flowers.*—Pink, clustered. *Calyx.*—Minute. *Corolla.*—Funnel-shaped, with five long recurved lobes. *Stamens.*—Five or ten, long, protruding noticeably. *Pistil.*—One, long, protruding.

Our May swamps and moist woods are made rosy by masses of the pink azalea which is often known as the wild honey-suckle, although not even a member of the Honeysuckle family. It is in the height of its beauty before the blooming of the laurel, and heralds the still lovelier pageant which is even then in rapid course of preparation.

In the last century the name of Mayflower was given to the

PLATE LXV

PINK AZALEA.—*R. nudiflorum.*

shrub by the Swedes in the neighborhood of Philadelphia. Peter Kalm, the pupil of Linnæus, after whom our laurel, *Kalmia*, is named, writes the following description of the shrub in his "Travels," which were published in English in 1771, and which explain the origin of one of its titles: "Some of the Swedes and Dutch call them Pinxter-bloem (Whitsunday-flower) as they really are in bloom about Whitsuntide; and at a distance they have some similarity to the Honeysuckle or 'Lonicera.' . . . Its flowers were now open and added a new ornament to the woods. . . . They sit in a circle round the stem's extremity and have either a dark red or a lively red color; but by standing for some time the sun bleaches them, and at last they get to a whitish hue. . . . They have some smell, but I cannot say it is very pleasant. However, the beauty of the flower entitles them to a place in every flower-garden." While our pink azalea could hardly be called "dark red" under any circumstances, it varies greatly in the color of its flowers.

The azalea is the national flower of Flanders.

——— ———

Rhododendron Rhodora. Heath Family.

A shrub from one to two feet high. *Leaves.*—Oblong, pale. *Flowers.*—Purplish-pink. *Calyx.*—Small. *Corolla.*—Two-lipped, almost without any tube. *Stamens.*—Ten, not protruding. *Pistil.*—One, not protruding.

> In May, when sea-winds pierced our solitudes,
> I found the fresh Rhodora in the woods,
> Spreading its leafless blooms in a damp nook,
> To please the desert and the sluggish brook.
> The purple petals, fallen in the pool,
> Made the black water with their beauty gay;
> Here might the red-bird come his plumes to cool,
> And court the flower that cheapens his array.
> Rhodora! if the sages ask thee why
> This charm is wasted on the earth and sky,
> Tell them, dear, that if eyes were made for seeing,
> Then Beauty is its own excuse for being;
> Why thou wert there, O rival of the rose!
> I never thought to ask, I never knew;
> But in my simple ignorance, suppose
> The self-same Power that brought me there, brought you. *

* Emerson.

SHEEP LAUREL. LAMBKILL.

Kalmia angustifolia. Heath Family.

A shrub from one to three feet high. *Leaves.*—Narrowly oblong, light green. *Flowers.*—Deep pink, in lateral clusters. *Calyx.*—Five-parted. *Corolla.*—Five-lobed, between wheel and bell-shaped, with stamens caught in its depressions as in the mountain laurel. *Stamens.*—Ten. *Pistil.*—One.

This low shrub grows abundantly with the mountain laurel, bearing smaller deep pink flowers at the same season, and narrower, paler leaves. It is said to be the most poisonous of the genus, and to be especially deadly to sheep, while deer are supposed to feed upon its leaves with impunity.

AMERICAN CRANBERRY.

Vaccinium macrocarpon. Heath Family.

Stems.—Slender, trailing, one to four feet long. *Leaves.*—Oblong, obtuse. *Flowers.*—Pale pink, nodding. *Calyx.*—With short teeth. *Corolla.*—Four-parted. *Stamens.*—Eight or ten, protruding. *Fruit.*—A large, acid, red berry.

In the peat-bogs of our Northeastern States we may look in June for the pink nodding flowers, and in late summer for the large red berries of this well-known plant.

ADDER'S MOUTH.

Pogonia ophioglossoides. Orchis Family (p. 17).

Stem.—Six to nine inches high, from a fibrous root. *Leaves.*—An oval or lance-oblong one near the middle of the stem, and a smaller or bract-like one near the terminal flower, occasionally one or two others, with a flower in their axils. *Flower.*—Pale pink, sometimes white, sweet-scented, one inch long, lip bearded and fringed.

Mr. Baldwin maintains that there is no wild flower of as pure a pink as this unless it be the *Sabbatia.* Its color has also been described as a " peach-blossom red." As already mentioned, the plant is found blossoming in bogs during the early summer in company with the Calopogons and sundews. Its violet-like fragrance greatly enhances its charm.

COMMON MILKWORT.

Polygala sanguinea. Milkwort Family.

Stem.—Six inches to a foot high, sparingly branched above, leafy to the top. *Leaves.*—Oblong-linear. *Flowers.*—Growing in round or oblong heads which are somewhat clover-like in appearance, bright pink or almost red, occasionally paler. *Calyx.*—Of five sepals, three of which are small and often greenish, while the two inner ones are much larger and colored like the petals. *Corolla.*—Of three petals connected with each other, the lower one keel-shaped. *Stamens.*—Six or eight. *Pistil.*—One. (Flowers too difficult to be analyzed by the non-botanist.)

· This pretty little plant abounds in moist and also sandy places, growing on mountain heights as well as in the salt meadows which skirt the sea. In late summer its bright flower-heads gleam vividly through the grasses, and from their form and color might almost be mistaken for pink clover. Occasionally they are comparatively pale and inconspicuous.

Polygala polygama. Milkwort Family.

Stems.—Very leafy, six to nine inches high, with cleistogamous flowers on underground runners. *Leaves.*—Lance-shaped or oblong. *Flowers.*—Purple-pink, loosely clustered in a terminal raceme. *Keel of Corolla.*—Crested. *Stamens.*—Eight. *Pistil.*—One.

Like its more attractive sister, the fringed polygala, this little plant hides its most useful, albeit unattractive, blossoms in the ground, where they can fulfil their destiny of perpetuating the species without danger of molestation by thievish insects or any of the distractions incidental to a more worldly career. Exactly what purpose the little above-ground flowers, which appear so plentifully in sandy soil in July, are intended to serve, it is difficult to understand.

FRINGED POLYGALA.

Polygala paucifolia. Milkwort Family.

Flowering-stems.—Three or four inches high, from long, prostrate or underground shoots which also bear cleistogamous flowers. *Leaves.*—The lower, small and scale-like, scattered, the upper, ovate, and crowded at the summit. *Flowers.*—Purple-pink, rarely white, rather large. *Keel of Corolla.*—Conspicuously fringed and crested. *Stamens.*—Six. *Pistil.*—One.

" I must not forget to mention that delicate and lovely flower of May, the fringed polygala. You gather it when you go for

PLATE LXVI

P. polygama.   *P. sanguinea.*

MILKWORT.

187

the fragrant showy orchis—that is, if you are lucky enough to find it. It is rather a shy flower, and is not found in every wood. One day we went up and down through the woods looking for it—woods of mingled oak, chestnut, pine, and hemlock, —and were about giving it up when suddenly we came upon a gay company of them beside an old wood-road. It was as if a flock of small rose-purple butterflies had alighted there on the ground before us. The whole plant has a singularly fresh and tender aspect. Its foliage is of a slightly purple tinge and of very delicate texture. Not the least interesting feature about the plant is the concealed fertile flower which it bears on a subterranean stem, keeping, as it were, one flower for beauty and one for use."

It seems unnecessary to tempt "odorous comparisons" by endeavoring to supplement the above description of Mr. Burroughs.

MOSS POLYGALA.

Polygala cruciata.   Milkwort Family.

Stems.—Three to ten inches high, almost winged at the angles, with spreading opposite leaves and branches.   *Leaves.*—Linear, nearly all whorled in fours.   *Flowers.*—Greenish or purplish-pink, growing in short, thick spikes which terminate the branches.

There is something very moss-like in the appearance of this little plant which blossoms in late summer. It is found near moist places and salt marshes along the coast, being very common in parts of New England.

SPREADING DOGBANE.   INDIAN HEMP.

Apocynum androsæmifolium.   Dogbane Family.

Stems.—Erect, branching, two or three feet high.   *Leaves.*—Opposite, oval.   *Flowers.*—Rose-color veined with deep pink, loosely clustered.   *Calyx.*—Five-parted.   *Corolla.*—Small, bell-shaped, five-cleft.   *Stamens.*—Five, slightly adherent to the pistil.   *Pistil.*—Two ovaries surmounted by a large, two-lobed stigma.   *Fruit.*—Two long and slender pods.

The flowers of the dogbane, though small and inconspicuous are very beautiful if closely examined. The deep pink veining of the corolla suggests nectar, and the insect-visitor is not mis-

PLATE LXVII

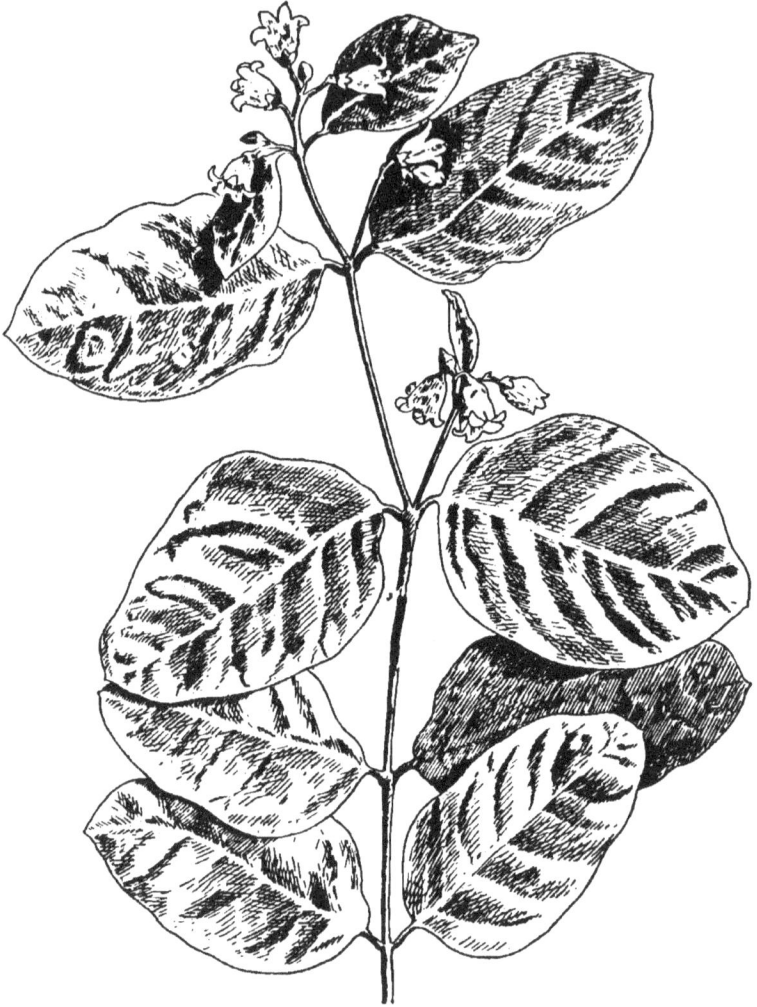

SPREADING DOGBANE.—*A. androsæmifolium.*

led, for at its base are five nectar-bearing glands. The two long, slender seed-pods which result from a single blossom seem inappropriately large, often appearing while the plant is still in flower. Rafinesque states that from the stems may be obtained a thread similar to hemp which can be woven into cloth, from the pods, cotton, and from the blossoms, sugar. Its generic and one of its English titles arose from the belief, which formerly prevailed, that it was poisonous to dogs. The plant is constantly found growing in roadside thickets, with bright, pretty foliage, and blossoms that appear in early summer.

HEDGE BINDWEED.

Convolvulus Americanus. Convolvulus Family.

Stem. — Twining or trailing. *Leaves.* — Somewhat arrow - shaped. *Flowers.*—Pink. *Calyx.*—Of five sepals enclosed in two broad leafy bracts. *Corolla.*—Five-lobed, bell-shaped. *Stamens.*—Five. *Pistil.*—One, with two stigmas.

Many an unsightly heap of rubbish left by the roadside is hidden by the delicate pink bells of the hedge bindweed, which again will clamber over the thickets that line the streams and about the tumbled stone-wall that marks the limit of the pasture. The pretty flowers at once suggest the morning-glory, to which they are closely allied.

The common European bindweed, *C. arvensis*, has white or pinkish flowers, without bracts beneath the calyx, and a low procumbent or twining stem. It has taken possession of many of our old fields where it spreads extensively and proves troublesome to farmers.

PURPLE-FLOWERING RASPBERRY.

Rubus odoratus. Rose Family.

Stem.—Shrubby, three to five feet high; branching, branches bristly and glandular. *Leaves.*—Three to five-lobed, the middle lobe prolonged. *Flowers.*—Purplish-pink, large and showy, two inches broad. *Calyx.*—Five-parted. *Corolla.*—Of five rounded petals. *Stamens and Pistils.*—Numerous. *Fruit.*—Reddish, resembling the garden raspberry.

This flower betrays its relationship to the wild rose, and might easily be mistaken for it, although a glance at the undi-

PLATE LXVIII

Fruit.

PURPLE-FLOWERING RASPBERRY.—*R. odoratus.*

vided leaves would at once correct such an error. The plant is a decorative one when covered with its showy blossoms, constantly arresting our attention along the wooded roadsides in June and July.

PALE CORYDALIS.

Corydalis glauca. Fumitory Family.

Stem.—Six inches to two feet high. *Leaves.*—Pale, divided into delicate leaflets. *Flowers.*—Pink and yellow, in loose clusters. *Calyx.*—Of two small, scale-like sepals. *Corolla.*—Pink, tipped with yellow; closed and flattened, of four petals, with a short spur at the base of the upper petal. *Stamens.*—Six, maturing before the pistil, thus avoiding self-fertilization. *Pistil.*—One.

From the rocky clefts in the summer woods springs the pale corydalis, its graceful foliage dim with a whitish bloom, and its delicate rosy, yellow-tipped flowers betraying by their odd flat corollas their kinship with the Dutchman's breeches and squirrel corn of the early year, as well as with the bleeding hearts of the garden. Thoreau assigns them to the middle of May, and says they are " rarely met with," which statement does not coincide with the experience of those who find the rocky woodlands each summer abundantly decorated with their fragile clusters.

The generic name, *Corydalis*, is the ancient Greek title for the crested lark, and said to refer to the crested seeds of this genus. The specific title, *glauca*, refers to the pallor of leaves and stem.

The golden corydalis, *C. aurea*, is found on rocky banks somewhat westward.

COMMON MILKWEED.

Asclepias Cornuti. Milkweed Family.

Stem.—Tall, stout, downy, with a milky juice. *Leaves.*—Generally opposite or whorled, the upper sometimes scattered, large, oblong, pale, minutely downy underneath. *Flowers.*—Dull, purplish-pink, clustered at the summit and along the sides of the stem. (These flowers are too difficult to be successfully analyzed by the non-botanist.) *Calyx.*—Five-parted, the divisions small and reflexed. *Corolla.*—Deeply five-parted, the divisions reflexed; above them a crown of five hooded nectaries, each containing an incurved horn. *Stamens.*—Five, inserted on the base of the corolla, united with each other and enclosing the pistils. *Pistils.*—Properly two, enclosed by the stamens, surmounted by a large five-angled disk. *Fruit.*—Two pods, one of which is large and full of silky-tufted seeds, the other often stunted.

This is probably the commonest representative of this striking and beautiful native family. The tall, stout stems, large, pale leaves, dull pink clustered flowers which appear in July, and later the puffy pods filled with the silky-tufted seeds beloved of imaginative children, are familiar to nearly everyone who spends a portion of the year in the country. The young sprouts are said to make an excellent pot-herb ; the silky hairs of the seed-pods have been used for the stuffing of pillows and mattresses, and can be mixed with flax or wool and woven to advantage ; while paper has been manufactured from the stout stalks.

The four-leaved milkweed, *A. quadrifolia*, is the most delicate member of the family, with fragrant rose-tinged flowers which appear on the dry wooded hill-sides quite early in June, and slender stems which are usually leafless below, and with one or two whorls and one or two pairs of oval, taper-pointed leaves above.

The swamp milkweed, *A. incarnata*, grows commonly in moist places. Its very leafy stems are two or three feet high, with narrowly oblong, pointed leaves. Its intense purple-pink flowers gleam from the wet meadows nearly all summer. They are smaller than those of the purple milkweed, *A. purpurascens*, which abounds in dry ground, and which may be classed among the deep pink or purple flowers according to the eye of the beholder.

HERB ROBERT.

Geranium Robertianum. Geranium Family.

Stem.—Forking, slightly hairy. *Leaves.*—Three, divided, the divisions again dissected. *Flowers.*—Purple-pink, small. *Calyx.*—Of five sepals. *Corolla.*—Of five petals. *Stamens.*—Ten. *Pistil.*—One, with five styles which split apart in fruit.

From June until October many of our shaded woods and glens are abundantly decorated by the bright blossoms of the herb Robert. The reddish stalks of the plant have won it the name of " red-shanks " in the Scotch Highlands. Its strong scent is caused by a resinous secretion which exists in several of the geraniums. In some species this resin is so-abundant that the stems will burn like torches, yielding a powerful and pleasant perfume. The common name is said to have been given the plant on account of its supposed virtue in a disease which was known as "Robert's plague," after Robert, Duke of Normandy. In some of the early writers it is alluded to as the "holy herb of Robert."

In fruit the styles of this plant split apart with an elasticity which serves to project the seeds to a distance, it is said, of twenty-five feet.

BUSH CLOVER.

Lespedeza procumbens. Pulse Family (p. 16).

Stems.—Slender, trailing, and prostrate. *Leaves.*—Divided into three clover-like leaflets. *Flowers.*—Papilionaceous, purplish-pink, veiny. *Pod.*—Small, rounded, flat, one-seeded.

The flowers of this plant often have the appearance of springing directly from the earth amid a mass of clover leaves. They are common in dry soil in the late summer and autumn, as are the other members of the same genus.

L. reticulata is an erect, very leafy species with similar blossoms, which are chiefly clustered near the upper part of the stem. The bush clovers betray at once their kinship with the tick-trefoils, but are usually found in more sandy, open places.

L. polystachya has upright wand-like stems from two to four feet high. Its flowers grow in oblong spikes on elongated stalks. Those of *L. capitata* are clustered in globular heads.

TICK-TREFOIL.

Desmodium Canadense. Pulse Family (p. 16).

Stem.—Hairy, three to six feet high. *Leaves.*—Divided into three somewhat oblong leaflets. *Flowers.*—Papilionaceous, dull purplish-pink, growing in densely flowered racemes. *Pod.*—Flat, deeply lobed on the lower margin, from one to three inches long, roughened with minute hooked hairs by means of which they adhere to animals and clothing.

Great masses of color are made by these flowers in the bogs and rich woods of midsummer. They are effective when seen in the distance, but rather disappointing on closer examination, and will hardly bear gathering or transportation. They are by far the largest and most showy of the genus.

PLATE LXIX

HERB ROBERT.—*G. Robertianum.*

195

TICK TREFOIL.

Desmodium nudiflorum. Pulse Family (p. 16).

Scape.—About two feet long. *Leaves.*—Divided into three broad leaflets, crowded at the summit of the flowerless stems. *Flowers.*—Papilionaceous, purplish-pink, small, growing in an elongated raceme on a mostly leafless scape.

This is a smaller, less noticeable plant than *D. Canadense.* It flourishes abundantly in dry woods, where it often takes possession in late summer to the exclusion of nearly all other flowers.

The flowers of *D. acuminatum* grow in an elongated raceme from a stem about whose summit the leaves, divided into very large leaflets, are crowded ; otherwise it resembles *D. nudiflorum.*

D. Dillenii grows to a height of from two to five feet, with erect leafy stems and medium-sized flowers. It is found commonly in open woods.

Many of us who do not know these plants by name have uttered various imprecations against their roughened pods. Thoreau writes : " Though you were running for your life, they would have time to catch and cling to your clothes. . . . These almost invisible nets, as it were, are spread for us, and whole coveys of desmodium and bidens seeds steal transportation out of us. I have found myself often covered, as it were, with an imbricated coat of the brown desmodium seeds or a bristling *chevaux-de-frise* of beggar-ticks, and had to spend a quarter of an hour or more picking them off in some convenient spot ; and so they get just what they wanted—deposited in another place."

BOUNCING BET. SOAPWORT.

Saponaria officinalis. Pink Family.

Stem.—Rather stout, swollen at the joints. *Leaves.*—Oval, opposite. *Flowers.*—Pink or white, clustered. *Calyx.*—Of five united sepals. *Corolla.*—Of five pinkish, long-clawed petals (frequently the flowers are double). *Stamens.*—Ten. *Pistil.*—One, with two styles.

A cheery pretty plant is this with large, rose-tinged flowers which are especially effective when double.

Bouncing Bet is of a sociable turn and is seldom found far from civilization, delighting in the proximity of farm-houses and their belongings, in the shape of children, chickens, and cattle.

PLATE LXX

BOUNCING BET.—*S. officinalis.*

197

She comes to us from England, and her "feminine comeliness and bounce" suggest to Mr. Burroughs a Yorkshire housemaid. The generic name is from *sapo*—soap, and refers to the lather which the juice forms with water, and which is said to have been used as a substitute for soap.

STEEPLE-BUSH.   HARDHACK.
Spiræa tomentosa. Rose Family.

Stems.—Very woolly. *Leaves.*—Alternate, oval, toothed. *Flowers.*—Small, pink, in pyramidal clusters. *Calyx.*—Five-cleft. *Corolla.*—Of five rounded petals. *Stamens.*—Numerous. *Pistils.*—Five to eight.

The pink spires of this shrub justify its rather unpoetic name of steeple-bush. It is closely allied to the meadow-sweet (Pl. XXVI.), blossoming with it in low grounds during the summer. It differs from that plant in the color of its flowers and in the woolliness of its stems and the lower surface of its leaves.

DEPTFORD PINK.
Dianthus Armeria. Pink Family.

One or two feet high. *Leaves.*—Opposite, long and narrow, hairy. *Flowers.*—Pink, with white dots, clustered. *Calyx.*—Five-toothed, cylindrical, with awl-shaped bracts beneath. *Corolla.*—Of five small petals. *Stamens.*—Ten. *Pistil.*—One, with two styles.

In July and August we find these little flowers in our eastern fields. The generic name, which signifies *Jove's own flower*, hardly applies to these inconspicuous blossoms. Perhaps it was originally bestowed upon *D. caryophyllus*, a large and fragrant English member of the genus, which was the origin of our garden carnation.

PURPLE LOOSESTRIFE.
Lythrum Salicaria. Loosestrife Family.

Stem.—Tall and slender. *Leaves.*—Lance-shaped, with a heart-shaped base, sometimes whorled in threes. *Flowers.*—Deep purple-pink, crowded and whorled in an interrupted spike. *Calyx.*—Five to seven-toothed, with little processes between the teeth. *Corolla.*—Of five or six somewhat wrinkled petals. *Stamens.*—Usually twelve, in two sets, six longer and six shorter. *Pistil.*—One, varying in size in the different blossoms, being of three different lengths.

One who has seen an inland marsh in August aglow with this beautiful plant, is almost ready to forgive the Old Country

PLATE LXXI

PURPLE LOOSESTRIFE.—*L. Salicaria.*

some of the many pests she has shipped to our shores in view of this radiant acquisition. The botany locates it anywhere between Nova Scotia and Delaware. It may be seen in the perfection of its beauty along the marshy shores of the Hudson and in the swamps of the Wallkill Valley.

When we learn that these flowers are called "long purples," by the English country people, the scene of Ophelia's tragic death rises before us :

> There is a willow grows aslant a brook,
> That shows his hoar leaves in the glassy stream,
> There with fantastic garlands did she come,
> Of crow-flowers, nettles, daisies, and long purples
> That liberal shepherds give a grosser name,
> But our cold maids do dead men's fingers call them.

Dr. Prior, however, says that it is supposed that Shakespeare intended to designate the purple-flowering orchis, *O. mascula*, which is said to closely resemble the showy orchis (Pl. LXII.) of our spring woods.

The flowers of the purple loosestrife are especially interesting to botanists on account of their *trimorphism*, which word signifies *occurring in three forms*, and refers to the stamens and pistils, which vary in size in the different blossoms, being of three different lengths, the pollen from any given set of stamens being especially fitted to fertilize a pistil of corresponding length.

MEADOW-BEAUTY. DEER-GRASS.

Rhexia Virginica. Melastoma Family.

Stem.—Square, with wing-like angles. *Leaves.*—Opposite, narrowly oval. *Flowers.*—Purplish-pink, clustered. *Calyx-tube.*—Urn-shaped, four-cleft at the apex. *Corolla.*—Of four large rounded petals. *Stamens.*—Eight, with long curved anthers. *Pistil.*—One.

It is always a pleasant surprise to happen upon a bright patch of these delicate deep-hued flowers along the marshes or in the sandy fields of midsummer. Their fragile beauty is of that order which causes it to seem natural that they should belong to a genus which is the sole northern representative of a tropical family. In parts of New England they grow in profusion, while in Arkansas the plant is said to be a great favorite with the deer,

PLATE LXXII

MEADOW-BEAUTY.—*R. Virginica.*

201

hence one of its common names. The flower has been likened to a scarlet evening primrose, and there is certainly a suggestion of the evening primrose in the four rounded, slightly heart-shaped petals. The protruding stamens, with their long yellow anthers, are conspicuous.

Of the plant in the late year, Thoreau writes : " The scarlet leaves and stem of the rhexia, sometime out of flower, make almost as bright a patch in the meadows now as the flowers did. Its seed-vessels are perfect little cream-pitchers of graceful form."

CLAMMY CUPHEA. WAX-WEED.

Cuphea viscosissima. Loosestrife Family.

Stem.—Sticky, hairy, branching. *Leaves.*—Usually opposite, rounded, lance-shaped. *Flowers.*—Deep purplish-pink, solitary or in racemes. *Calyx.*—Tubular, slightly spurred at the base on the upper side, six-toothed at the apex, usually with a slight projection between each tooth. *Corolla.*—Small, of six unequal petals. *Stamens.*—Eleven or twelve, of unequal sizes, in two sets. *Pistil.*—One, with a two-lobed stigma.

In the dry fields and along the roadsides of late summer this plant is found in blossom. Its rather wrinkled purplish-pink petals and unequal stamens suggest the flowers of the spiked loosestrife, *L. Salicaria*, to which it is closely related.

SEA PINK.

Sabbatia stellaris. Gentian Family.

Stem.—Slender, loosely branched. *Leaves.*—Opposite, oblong to lance-shaped, the upper narrowly linear. *Flowers.*—Large, deep pure pink to almost white. *Calyx.*—Usually five-parted, the lobes long and slender. *Corolla.*—Usually five-parted, conspicuously marked with red and yellow in the centre. *Stamens.*—Usually five. *Pistil.*—One, with two-cleft style.

The advancing year has few fairer sights to show us than a salt meadow flushed with these radiant blossoms. They are so abundant, so deep-hued, so delicate ! One feels tempted to lie down among the pale grasses and rosy stars in the sunshine of the August morning and drink his fill of their beauty. How often nature tries to the utmost our capacity of appreciation and leaves us still insatiate ! At such times it is almost a relief to turn from the mere contemplation of beauty to the study of its structure ; it rests our overstrained faculties.

PLATE LXXIII

SEA PINK.—*S. stellaris.*

The vivid coloring and conspicuous marking of these flowers indicate that they aim to attract certain members of the insect world. As in the fireweed the pistil of the freshly opened blossom is curved sideways, with its lobes so closed and twisted as to be inaccessible on their stigmatic surfaces to the pollen which the already mature stamens are discharging. When the effete anthers give evidence'that they are *hors de combat* by their withered appearance, the style erects itself and spreads its stigmas.

S. angularis is a species which may be found in rich soil inland. Its somewhat heart-shaped, clasping, five-nerved leaves and angled stem serve to identify it.

S. chloroides is a larger and peculiarly beautiful species which borders brackish ponds along the coast. Its corolla is about two inches broad and eight to twelve-parted.

MARSH ST. JOHN'S-WORT.

Elodes campanulata. St. John's-wort Family.

Stem.—One or two feet high, often pinkish, later bright red. *Leaves.*— Opposite, set close to the stem or clasping by a broad base. *Flowers.*— Pinkish or flesh-color, small, closely clustered at the summit of the stem and in the axils of the leaves. *Calyx.*—Of five sepals, often pinkish. *Corolla.* —Of five petals. *Stamens.*—Nine, in three sets, the sets separated by orange-colored glands. *Pistil.*—One, with three styles.

If one has been so unlucky, from the usual point of view, or so fortunate, looking at the matter with the eyes of the flower-lover, as to find himself in a rich marsh early in August, his eye is likely to fall upon the small, pretty pinkish flowers and pale clasping leaves of the marsh St. John's-wort. A closer inspection will discover that the foliage is dotted with the pellucid glands, and that the stamens are clustered in groups after the family fashion. Should the same marsh be visited a few weeks later dashes of vivid color will guide one to the spot where the little pink flowers were found. In their place glow the conspicuous ovaries and bright leaves which make the plant very noticeable in late August.

Elodes is a corruption from a Greek word which signifies *growing in marshes.*

PLATE LXXIV

Sabbatia chloroides.

205

ROSE MALLOW. SWAMP MALLOW.

Hibiscus Moscheutos. Mallow Family.

Stem.—Stout and tall, four to eight feet high. *Leaves.*—The lower three-lobed, the upper oblong, whitish and downy beneath. *Flowers.*—Large and showy, pink. *Calyx.*—Five-cleft, with a row of narrow bractlets beneath. *Corolla.*—Of five large petals. *Stamens.*—Many, on a tube which encloses the lower part of the style. *Pistils.*—Five, united into one, with five stigmas which are like pin-heads.

When the beautiful rose mallow slowly unfolds her pink banner-like petals and admits the eager bee to her stores of golden pollen, then we feel that the summer is far advanced. As truly as the wood anemone and the blood-root seem filled with the essence of spring and the promise of the opening year, so does this stately flower glow with the maturity and fulfilment of late summer. Here is none of the timorousness of the early blossoms which peep shyly out, as if ready to beat a hasty retreat should a late frost overtake them, but rather a calm assurance that the time is ripe, and that the salt marshes and brackish ponds are only awaiting their rosy lining.

The marsh mallow, whose roots yield the mucilaginous substance utilized in the well-known confection, is *Althæa officinalis,* an emigrant from Europe. It is a much less common plant than the *Hibiscus,* its pale pink flowers being found in some of the salt marshes of New England and New York.

The common mallow, *Malva rotundifolia,* which overruns the country dooryards and village waysides, is a little plant with rounded, heart-shaped leaves and small purplish flowers. It is used by the country people for various medicinal purposes and is cultivated and commonly boiled with meat in Egypt. Job pictures himself as being despised by those who had been themselves so destitute as to " cut up mallows by the bushes. . . . for their meat." *

* Job xxx. 4.

PLATE LXXV

ROSE MALLOW.—*H. Moscheutos.*

207

SALT MARSH FLEABANE.

Pluchea camphorata. Composite Family (p. 13).

Stem.—Two to five feet high. *Leaves.*—Pale, thickish, oblong or lance-shaped, toothed. *Flower-heads.*—Pink, small, in flat-topped clusters, composed entirely of tubular flowers.

In the salt marshes where we find the starry sea pinks and the feathery sea lavender, we notice a pallid-looking plant whose pink flower-buds are long in opening. It is late summer or autumn before the salt marsh fleabane is fairly in blossom. There is a strong fragrance to the plant which hardly suggests camphor, despite its specific title.

HAIRY WILLOW-HERB.

Epilobium hirsutum. Evening Primrose Family.

Three to five feet high. *Stem.*—Densely hairy, stout, branching. *Leaves.*—Mostly opposite, lance-oblong, finely toothed. *Flower.*—Purplish, pink, small, in the axils of the upper leaves, or in a leafy, short raceme. *Calyx.*—Four or five-parted. *Corolla.*—Of four petals. *Stamens.*—Eight. *Pistil.*—One, with a four-parted stigma.

The hairy willow-herb is found in waste places, blossoming in midsummer. It is an emigrant from Europe.

FIREWEED. GREAT WILLOW-HERB.

Epilobium angustifolium. Evening Primrose Family.

Stem.—Four to seven feet high. *Leaves.*—Scattered, lance-shaped, willow-like. *Flowers.*—Purplish-pink, large, in a long raceme. *Calyx.*—Four-cleft. *Corolla.*—Of four petals. *Stamens.*—Eight. *Pistil.*—One, with a four-lobed stigma.

In midsummer this striking plant begins to mass its deep-hued blossoms along the roadsides and low meadows. It is supposed to flourish with especial abundance in land that has newly been burned over; hence, its common name of fireweed. Its willow-like foliage has given it its other English title. The likeness between the blossoms of this plant and those of the evening primrose betray their kinship. When the stamens of the fireweed first mature and discharge their pollen the still immature style is curved backward and downward with its stigmas closed. Later it straightens and lengthens to its full dimensions,

PLATE LXXVI

FIREWEED.—*E. angustifolium.*

209

so spreading its four stigmas as to be in position to receive the pollen of another flower from the visiting bee.

Epilobium coloratum. Evening Primrose Family.

One to three feet high. *Leaves.*—Rather large, lance-shaped, sharply toothed. *Flowers.*—Pale pink, small, more or less nodding, resembling in structure those of the hairy willow-herb. *Pistil.*—One, with a club-shaped stigma.

This species is found in abundance in wet places in summer.

PURPLE GERARDIA.

Gerardia purpurea. Figwort Family.

Stem.—One to four feet high, widely branching. *Leaves.*—Linear, sharply pointed. *Flowers.*—Bright purplish-pink, rather large. *Calyx.*—Five-toothed. *Corolla.*—One inch long, somewhat tubular, swelling above, with five more or less unequal, spreading lobes, often downy and spotted within. *Stamens.*—Four, in pairs, hairy. *Pistil.*—One.

In late summer and early autumn these pretty noticeable flowers brighten the low-lying ground along the coast and in the neighborhood of the Great Lakes. The sandy fields of New England and Long Island are oftentimes a vivid mass of color owing to their delicate blossoms. The plant varies somewhat in the size of its flowers and in the manner of its growth.

The little seaside gerardia, *G. maritima*, is from four inches to a foot high. Its smaller blossoms are also found in salt marshes.

The slender gerardia, *G. tenuifolia*, is common in mountainous regions. The leaves of this species are exceedingly narrow. Like the false foxglove (Pl. LIX.) and other members of this genus, these plants are supposed to be parasitic in their habits.

JOE-PYE-WEED. TRUMPET-WEED.

Eupatorium purpureum. Composite Family (p. 13).

Stem.—Stout and tall, two to twelve feet high, often dotted. *Leaves.*—In whorls of three to six, oblong or oval, pointed, rough, veiny, toothed. *Flower-heads.*—Purplish-pink, small, composed entirely of tubular blossoms, with long protruding styles, growing in large clusters at or near the summit of the stem.

The summer is nearly over when the tall, conspicuous Joe-Pye-weed begins to tinge with " crushed raspberry " the lowlands

PLATE LXXVII

JOE-PYE-WEED.—*E. purpureum.*

211

through which we pass. In parts of the country it is nearly as common as the golden-rods and asters which appear at about the same season. With the deep purple of the iron-weed it gives variety to the intense hues which herald the coming of autumn.

"Joe Pye" is said to have been the name of an Indian who cured typhus fever in New England by means of this plant. The tiny trumpet-shaped blossoms which make up the flower-heads may have suggested the other common name.

PINK KNOTWEED.

Polygonum Pennsylvanicum. Buckwheat Family.

One to four feet high. *Stem.*—Branching. *Leaves.*—Alternate, lance-shaped. *Flowers.*— Bright pink, growing in thick, short, erect spikes. *Calyx.* — Mostly five-parted, the divisions petal-like, pink. *Corolla.*—None. *Stamens.*—Usually eight. *Pistil.*—One, with a two-cleft style.

In late summer this plant can hardly escape notice. Its erect pink spikes direct attention to some neglected corner in the garden or brighten the fields and roadsides. The rosy divisions of the calyx persist till after the fruit has formed, pressing closely against the dark seed-vessel within.

SAND KNOTWEED.

Polygonella articulata. (Formerly *Polygonum articulatum.*) Buckwheat Family.

Erect, branching, four to twelve inches high. *Leaves.*—Linear, inconspicuous. *Flowers.*—Rose-color, nodding, in very slender racemes, *Calyx.*—Five-parted. *Corolla.*—None. *Stamens.*—Eight. *Pistil.*—One, with three styles.

Under date of September 26th, Thoreau writes: "The *Polygonum articulatum,* giving a rosy tinge to Jenny's desert, is very interesting now, with its slender dense racemes of rose-tinted flowers, apparently without leaves, rising cleanly out of the sand. It looks warm and brave, a foot or more high, and mingled with deciduous blue curls. It is much divided, with many spreading, slender-racemed branches, with inconspicuous linear leaves, reminding me, both by its form and its colors, of a peach-orchard in blossom, especially when the sunlight falls on it ; minute rose-tinted flowers that brave the frosts, and advance the summer into fall, warming with their color sandy hill-sides and deserts, like the glow of evening reflected on the sand, apparently all flower and no leaf. Rising apparently with clean bare stems from the sand, it spreads out into this graceful head of slender, rosy racemes, wisp-like. This little desert of less than an acre blushes with it.''

NOTE.—The Moss Pink, *Phlox subulata*, with purple-pink flowers, and *Phlox glaberrima*, with pink or whitish flowers, will be found in the Blue and Purple section (p. 235). The Mountain Laurel (p. 57) and the American Rhododendron (p. 60) are frequently found bearing pink flowers. At times it has been difficult to determine whether certain flowers should be described in the Pink or in the Purple section. The reader should bear this in mind, consulting both in dubious cases.

IV

RED

WILD COLUMBINE.

Aquilegia Canadensis. Crowfoot Family.

Twelve to eighteen inches high. *Stem.*—Branching. *Leaves.*—Much-divided, the leaflets lobed. *Flowers.*—Large, bright red, yellow within, nodding. *Calyx.*—Of five red petal-like sepals. *Corolla.*—Of five petals in the form of large hollow spurs, which are red without and yellow within. *Stamens.*—Numerous. *Pistils.*—Five, with slender styles.

> —A woodland walk,
> A quest of river-grapes, a mocking thrush,
> A wild-rose or rock-loving columbine,
> Salve my worst wounds,

declares Emerson ; and while perhaps few among us are able to make so light-hearted and sweeping a claim for ourselves, yet many will admit the soothing power of which the woods and fields know the secret, and will own that the ordinary annoyances of life may be held more or less in abeyance by one who lives in close sympathy with nature.

About the columbine there is a daring loveliness which stamps it on the memories of even those who are not ordinarily minute observers. It contrives to secure a foothold in the most precipitous and uncertain of nooks, its jewel-like flowers gleaming from their lofty perches with a graceful *insouciance* which awakens our sportsmanlike instincts and fires us with the ambition to equal it in daring and make its loveliness our own. Perhaps it is as well if our greediness be foiled and we get a tumble for our pains, for no flower loses more with its surroundings than the columbine. Indeed, these destructive tendencies which are strong within most of us generally defeat themselves by decreasing our pleasure in a blossom the moment we have ruthlessly and without purpose snatched it from its environment. If we

PLATE LXXVIII

Fruit.

WILD COLUMBINE.—*A. Canadensis.*

215

honestly wish to study its structure, or to bring into our homes for preservation a bit of the woods' loveliness, its interest and beauty are sure to repay us.   But how many pluck every strik-ing flower they see only to toss it carelessly aside when they reach their destination, if they have not already dropped it by the way.   Surely if in such small matters sense and self-control were inculcated in children, more would grow up to the poet's standard of worthiness :

> Hast thou named all the birds without a gun ?
> Loved the wood-rose and left it on its stalk ? .
> At rich men's tables eaten bread and pulse?
> Unarmed, faced danger with a heart of trust *
> And loved so well a high behavior,
> In man or maid, that thou from speech refrained,
> Nobility more nobly to repay ?
> O, be my friend, and teach me to be thine ! *

The name of columbine is derived from *colomba*—a dove, but its significance is disputed.   Some believe that it was associated with the bird-like claws of the blossom ; while Dr. Prior main-tains that it refers to the " resemblance of its nectaries to the heads of pigeons in a ring around a dish, a favorite device of ancient artists."

The meaning of the generic title is also doubtful.   Gray de-rives it from *aquilegus*—water-drawing, but gives no further ex-planation, while other writers claim that it is from *aquila*, an eagle, seeing a likeness to the talons of an eagle in the curved nectaries.

WAKE ROBIN.   BIRTHROOT.

Trillium erectum.   Lily Family.

Stem.—Stout, from a tuber-like rootstock.   *Leaves.*—Broadly ovate, three in a whorl a short distance below the flower.   *Flower.*—Single, terminal, usually purplish-red, occasionally whitish, pinkish, or greenish, on an erect or somewhat inclined flower-stalk.   *Calyx.*—Of three green spreading se-pals.   *Corolla.*—Of three large lance-shaped petals.   *Stamens.*—Six.   *Pis-til.*—One, with three large spreading stigmas.   *Fruit.*—A large, ovate, six-angled reddish berry.

This wake robin is one of the few self-assertive flowers of the early year.   Its contemporaries act as if somewhat uncertain as

* Emerson.

PLATE LXXIX

Fruit.

WAKE ROBIN.—*T. erectum.*

217

to whether the spring had really come to stay, but no such lack of confidence possesses our brilliant young friend, who almost flaunts her lurid petals in your face, as if to force upon you the welcome news that the time of birds and flowers is at hand. Pretty and suggestive as is the common name, it is hardly appropriate, as the robins have been on the alert for many days before our flower unfurls its crimson signal. Its odor is most unpleasant. Its reddish fruit is noticeable in the woods of late summer.

The sessile trillium, *T. sessile*, has no separate flower-stalk, its red or greenish blossom being set close to the stem leaves. Its petals are narrower, and its leaves are often blotched or spotted. Its berry is globular, six-angled, and red or purplish.

The wake robins are native to North America, only one species being found just beyond the boundaries in the Russian territory.

WOOD BETONY. LOUSEWORT.

Pedicularis Canadensis. Figwort Family.

Stems.—Clustered, five to twelve inches high. *Leaves.*—The lower ones deeply incised, the upper less so. *Flowers.*—Yellow and red, growing in a short dense spike. *Calyx.*—Of one piece split in front. *Corolla.*—Two-lipped, the narrow upper lip arched, the lower three-lobed. *Stamens.*—Four. *Pistil.*—One.

The bright flowers of the wood betony are found in our May woods, often in the company of the columbine and yellow violet. Near Philadelphia they are said to be among the very earliest of the flowers, coming soon after the trailing arbutus. In the later year the plant attracts attention by its uncouth spikes of brown seed-pods.

Few wayside weeds have been accredited with greater virtue than the ancient betony, which a celebrated Roman physician claimed could cure forty-seven different disorders. The Roman proverb, "Sell your coat and buy betony," seems to imply that the plant did not flourish so abundantly along the Appian Way as it does by our American roadsides. Unfortunately we are reluctantly forced to believe once more that our native flower is not identical with the classic one, but that it has received its common name through some superficial resemblance to the original betony or *Betonica*.

PAINTED CUP.

Castilleia coccinea. Figwort Family.

Stem.—Hairy, six inches to a foot high. *Root-leaves.*—Clustered, oblong. *Stem-leaves.*—Incised, those among the flowers three to five-cleft, bright scarlet toward the summit, showy. *Flowers.*—Pale yellow, spiked. *Calyx.* —Tubular, flattened. *Corolla.*—Two-lipped, its upper lip long and narrow, its lower short and three-lobed. *Stamens.*—Four, unequal. *Pistil.*—One.

> ——Scarlet tufts
> Are glowing in the green like flakes of fire ;
> The wanderers of the prairie know them well,
> And call that brilliant flower the painted cup.*

But we need not go to the prairie in order to see this plant, for it is equally abundant in certain low sandy New England meadows as well as in the near vicinity of New York City. Under date of June 3d, Thoreau graphically describes its appearance near Concord, Mass. : "The painted cup is in its prime. It reddens the meadow, painted-cup meadow. It is a splendid show of brilliant scarlet, the color of the cardinal flower, and surpassing it in mass and profusion. . . . I do not like the name. It does not remind me of a cup, rather of a flame when it first appears. It might be called flame flower, or scarlet tip. Here is a large meadow full of it, and yet very few in the town have ever seen it. It is startling to see a leaf thus brilliantly painted, as if its tip were dipped into some scarlet tincture, surpassing most flowers in intensity of color."

WOOD LILY. WILD RED LILY.

Lilium Philadelphicum. Lily Family.

Stem.—Two to three feet high. *Leaves.*—Whorled or scattered, narrowly lance-shaped. *Flower.*—Erect, orange-red or scarlet, spotted with purple. *Perianth.*—Of six erect narrowly clawed sepals, with nectar-bearing furrows at their base. *Stamens.*—Six. *Pistil.*—One, with three-lobed stigma.

Here and there in the shadowy woods is a vivid dash of color made by some wild red lily which has caught a stray sunbeam

* Bryant.

in its glowing cup. The purple spots on its sepals guide the greedy bee to the nectar at their base ; we too can take the hint and reap a sweet reward if we will, after which we are more in sympathy with those eager, humming bees.

This erect, deep-hued flower is so different from its nodding sister of the meadows, that we wonder that the two should be so often confused. When seen away from its surroundings it has less charm perhaps than either the yellow or the Turk's-cap lily ; but when it rears itself in the cool depths of its woodland home we feel the uniqueness of its beauty.

TURK'S CAP LILY.

Lilium superbum. Lily Family.

Stem.—Three to seven feet high. *Leaves.*—Lance-shaped, the lower whorled. *Flowers.*—Orange or scarlet, with purple spots within, three inches long, from three to forty growing in pyramidal clusters. *Perianth.*—Of six strongly recurved sepals. *Stamens.*—Six, with long anthers. *Pistil.*—One, with a three-lobed stigma.

> Consider the lilies of the field, how they grow ;
> They toil not, neither do they spin ;
> And yet I say unto you, that even Solomon in all his glory
> Was not arrayed like one of these.

How they come back to us, the beautiful hackneyed lines, and flash into our memories with new significance of meaning when we chance suddenly upon a meadow bordered with these the most gorgeous of our wild flowers.

We might doubt whether our native lilies at all resembled those alluded to in the scriptural passage, if we did not know that a nearly allied species grew abundantly in Palestine ; for we have reason to believe that *lily* was a title freely applied by many Oriental poets to any beautiful flower.

Perhaps this plant never attains far inland the same luxuriance of growth which is common to it in some of the New England lowlands near the coast. Its radiant, nodding blossoms are seen in great profusion as we travel by rail from New York to Boston.

PLATE LXXX

WOOD LILY.—*L. Philadelphicum.*

221

HOUND'S TONGUE.

Cynoglossum officinale.   Borage Family.

Stem.—Clothed with soft hairs. *Leaves.*—Alternate, hairy, the upper
ones lance-shaped, clasping somewhat by a rounded or heart-shaped base.
Flowers.— Purplish-red, growing in a curved raceme-like cluster which
straightens as the blossoms expand. *Calyx.*—Five-parted. *Corolla*.—
Funnel-form, five-lobed. *Stamens.*—Five. *Pistil.*—One. *Fruit.*—A large
nutlet roughened with barbed or hooked prickles.

This coarse plant, whose disagreeable odor strongly suggests
mice, is not only a troublesome weed in pasture-land but a
special annoyance to wool-growers, as its prickly fruit adheres
with pertinacity to the fleece of sheep.   Its common name is a
translation of its generic title and refers to the shape and texture
of the leaves.   The dull red flowers appear in summer.

BUTTERFLY-WEED.   PLEURISY-ROOT.

Asclepias tuberosa.   Milkweed Family.

Stem.—Rough and hairy, one or two feet high, erect, very leafy, branching at the summit, without milky juice. *Leaves.*—Linear to narrowly lance-
shaped. *Flowers.*— Bright orange-red, in flat-topped, terminal clusters,
otherwise closely resembling those of the common milkweed (p. 192.) *Fruit.*
—Two hoary erect pods, one of them often stunted.

Few if any of our native plants add more to the beauty of the
midsummer landscape than the milkweeds, and of this family no
member is more satisfying to the color-craving eye than the
gorgeous butterfly-weed, whose vivid flower-clusters flame from
the dry sandy meadows with such luxuriance of growth as to
seem almost tropical.   Even in the tropics one hardly sees any-
thing more brilliant than the great masses of color made by
these flowers along some of our New England railways in July,
while farther south they are said to grow even more profusely.   Its gay coloring has given the plant its name of butterfly-
weed, while that of pleurisy-root arose from the belief that
the thick, deep root was a remedy for pleurisy.   The Indians
used it as food and prepared a crude sugar from the flowers; the
young seed-pods they boiled and ate with buffalo-meat.   The
plant is worthy of cultivation and is easily transplanted, as the
fleshy roots when broken in pieces form new plants.   Oddly

PLATE LXXXI

Fruit.

BUTTERFLY-WEED.—*A. tuberosa.*

223

enough, at the Centennial much attention was attracted by a bed of these beautiful plants which were brought from Holland. Truly, flowers like prophets are not without honor save in their own country.

EUROPEAN HAWKWEED. DEVIL'S PAINTBRUSH.

Hieracium aurantiacum. Composite Family (p. 13).

Stem.—Hairy, erect. *Leaves.*—Hairy, oblong, close to the ground. *Flower-heads.* — Orange-red, composed entirely of strap - shaped flowers, clustered.

In parts of New York and of New England the midsummer meadows are ablaze with the brilliant orange-red flowers of this striking European weed. It is among the more recent emigrants to this country and bids fair to become an annoyance to the farmer, hence its not altogether inappropriate title of devil's paintbrush. In England it was called " Grimm the Collier," on account of its black hairs and after a comedy of the same title which was popular during the reign of Queen Elizabeth. Both its common and generic names refer to an ancient superstition to the effect that birds of prey used the juice of this genus to strengthen their eyesight.

OSWEGO TEA. BEE BALM.

Monarda didyma. Mint Family (p. 16).

Stem.—Square, erect, about two feet high. *Leaves.*—Opposite, ovate, pointed, aromatic ; those near the flowers tinged with red. *Flowers.*— Bright red, clustered in a close round head. *Calyx.*—Reddish, five-toothed. *Corolla.* — Elongated, tubular, two-lipped. *Stamens.*—Two, elongated, protruding. *Pistil.*—One, with a two-lobed style, protruding.

We have so few red flowers that when one flashes suddenly upon us it gives us a pleasant thrill of wonder and surprise. Then red flowers know so well how to enhance their beauty by seeking an appropriate setting. They select the rich green backgrounds only found in moist, shady places, and are peculiarly charming when associated with a lonely marsh or a mountain brook. The bee balm especially haunts these cool nooks, and

PLATE LXXXII

Single flower.

OSWEGO TEA.—*M. didyma.*

225

its rounded flower-clusters touch with warmth the shadows of the damp woods of midsummer. The Indians named the flower *O-gee-chee*—flaming flower, and are said to have made a tea-like decoction from the blossoms.

PIMPERNEL. POOR-MAN'S-WEATHER-GLASS.

Anagallis arvensis. Primrose Family.

Stems.—Low, spreading. *Leaves.*—Opposite, ovate, set close to the stem. *Flowers.*—Red, occasionally blue or white, growing singly from the axils of the leaves. *Calyx.*—Five-parted. *Corolla.*—Five-parted, wheel-shaped. *Stamens.*—Five, with bearded filaments. *Pistil.*—One.

This flower is found in sandy fields, being noted for its sensitiveness to the weather. It folds its petals at the approach of rain, and fails to open at all on a wet or cloudy day. Even in fine weather it closes in the early afternoon and " sleeps " till the next morning. Its ripened seeds are of value as food for many song-birds. It was thought at one time to be serviceable in liver complaints, which reputed virtue may have given rise to the old couplet :

> No ear hath heard, no tongue can tell
> The virtues of the pimpernel.

CARDINAL-FLOWER.

Lobelia cardinalis. Lobelia Family.

Stem.—From two to four feet high. *Leaves.*—Alternate, narrowly oblong, slightly toothed. *Flowers.*—Bright red, growing in a raceme. *Calyx.*—Five-cleft. *Corolla.*—Somewhat two-lipped, the upper lip of two rather erect lobes, the lower spreading and three-cleft. *Stamens.*—Five, united into a tube. *Pistil.*—One, with a fringed stigma.

We have no flower which can vie with this in vivid coloring. In late summer its brilliant red gleams from the marshes or is reflected from the shadowy water's edge with unequalled intensity—

> As if some wounded eagle's breast
> Slow throbbing o'er the plain,
> Had left its airy path impressed
> In drops of scarlet rain.*

The early French Canadians were so struck with its beauty that they sent the plant to France as a specimen of what the wilds of

---

* Holmes.

PLATE LXXXIII

CARDINAL-FLOWER.—*L. cardinalis.*

the New World could yield. Perhaps at that time it received its English name which likens it to the gorgeously attired dignitaries of the Roman Church.

TRUMPET HONEYSUCKLE.

Lonicerà sempervirens. Honeysuckle Family.

A twining shrub. *Leaves.*—Entire, opposite, oblong, the upper pairs united around the stem. *Flowers.*—Deep red without, yellowish within ; in close clusters from the axils of the upper leaves. *Calyx.*—With very short teeth. *Corolla.*—Trumpet-shaped, five-lobed. *Stamens.*—Five. *Pistil.*—One. *Fruit.*—A red or orange berry.

Many of us are so familiar with these flowers in our gardens that we have, perhaps, considered them "escapes" when we found them brightening the pasture thicket where really they are most at home, appearing at any time from May till October.

The fragrant woodbine, *L. grata,* is also frequently cultivated. Its natural home is the rocky woodlands, where its sweet-scented whitish or yellowish flowers appear in May. Its stamens and style protrude conspicuously beyond the corolla-tube, which is an inch in length.

The greenish or yellowish flowers of the fly honeysuckle, *L. ciliata,* grow in pairs. They are found in the rocky woods of May, on an erect, bushy shrub, the leaves of which are all distinct, never meeting about the stem.

V

BLUE AND PURPLE

LIVERWORT. LIVER-LEAF.

Hepatica triloba. Crowfoot Family.

Scape.—Fuzzy, one-flowered. *Leaves.*—Rounded, three-lobed, from the root. *Flowers.*—Blue, white, or pinkish. *Calyx.*—Of six to twelve petal-like sepals ; easily taken for a corolla, because directly underneath are three little leaves which resemble a calyx. *Corolla.*—None. *Stamens.*—Usually numerous. *Pistils.*—Several.

> The liver-leaf puts forth her sister blooms
> Of faintest blue—

soon after the late snows have melted. Indeed these fragile-looking, enamel-like flowers are sometimes found actually beneath the snow, and form one of the many instances which we encounter among flowers, as among their human contemporaries, where the frail and delicate-looking withstand storm and stress far better than their more robust-appearing brethren. We welcome these tiny newcomers with especial joy, not alone for their delicate beauty, but because they are usually the first of all the flowers upon the scene of action, if we rule out the never-tardy skunk-cabbage. The rusty leaves of last summer are obliged to suffice for the plant's foliage until some little time after the blossoms have appeared, when the young fresh leaves begin to uncurl themselves. Some one has suggested that the fuzzy little buds look as though they were still wearing their furs as a protection against the wintry weather which so often stretches late into our spring. The flowers vary in color from a lovely blue to pink or white. They are found chiefly in the woods, but occasionally on the sunny hill-sides as well.

The generic name, *Hepatica*, is from the Greek for liver, and was probably given to the plant on account of the shape of its leaf. Dr. Prior says that "in consequence of this fancied likeness it was used as a remedy for liver-complaints, the common people having long labored under the belief that nature indicated in some such fashion the uses to which her creations might be applied."

COMMON BLUE VIOLET.

Viola cucullata. Violet Family.

Scape.—Slender, one - flowered. *Leaves.*—Heart - shaped, all from the root. *Flowers.*—Varying from a pale blue to deep purple, borne singly on a scape. *Calyx.*—Of five sepals extended into ears at the base. *Corolla.*—Of five somewhat unequal petals, the lower one spurred at the base. *Stamens.*—Short and broad, somewhat united around the pistil. *Pistil.*—One with a club-shaped style and bent stigma.

Perhaps this is the best-beloved as well as the best-known of the early wild flowers. Whose heart has not been gladdened at one time or another by a glimpse of some fresh green nook in early May where

—purple violets lurk,
With all the lovely children of the shade ?

It seems as if no other flower were so suggestive of the dawning year, so associated with the days when life was full of promise. Although I believe that more than a hundred species of violets have been recorded, only about thirty are found in our country ; of these perhaps twenty are native to the Northeastern States. Unfortunately we have no strongly sweet-scented species, none

—sweeter than the lids of Juno's eyes
Or Cytherea's breath,—

as Shakespeare found the English blossom. Prophets and warriors as well as poets have favored the violet ; Mahomet preferred it to all other flowers, and it was chosen by the Bonapartes as their emblem.

Perhaps its frequent mention by ancient writers is explained by the discovery that the name was one applied somewhat indiscriminately to sweet-scented blossoms.

The bird-foot violet, *V. pedata*, unlike other members of the family, has leaves which are divided into linear lobes. Its flower is peculiarly lovely, being large and velvety. The variety, *V. bi-*

PLATE LXXXIV

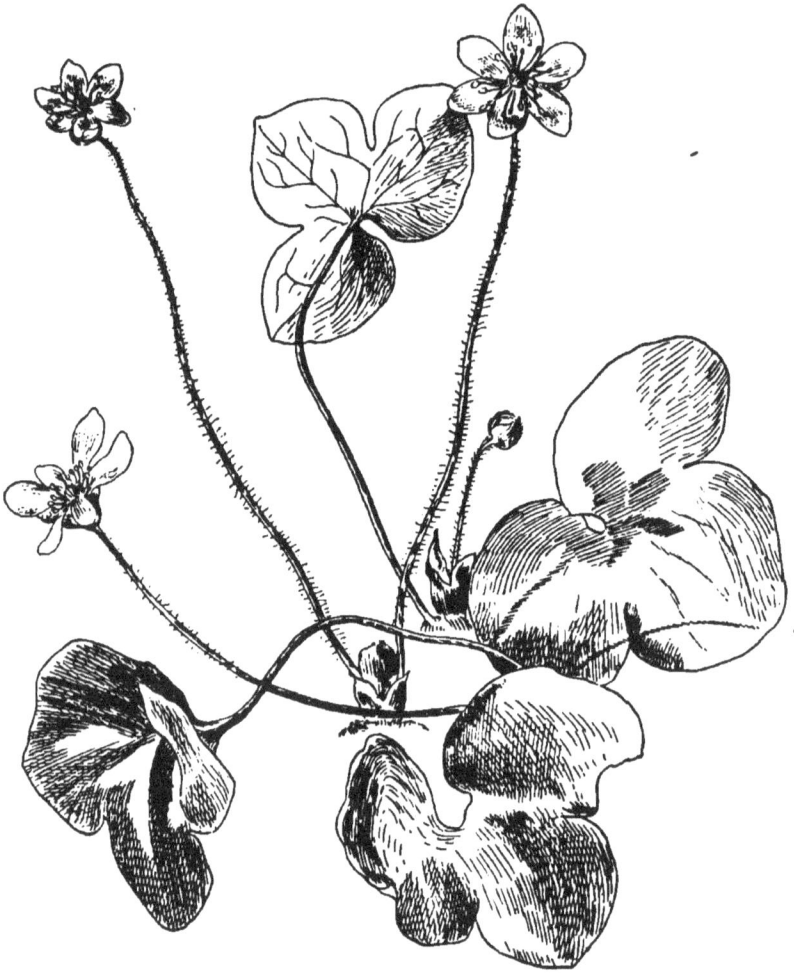

LIVERWORT.—*H. triloba.*

color, is especially striking and pansy-like, its two upper petals being of a deeper hue than the others. It is found in the neighborhood of Washington in abundance, and on the shaly soil of New Jersey.

An interesting feature of many of these plants is their cleistogamous flowers. These are small and inconspicuous blossoms, which never open(thus guarding their pollen against all depredations), but which are self-fertilized, ripening their seeds in the dark. They are usually found near or beneath the ground, and are often taken for immature buds.

DOG VIOLET.

Viola canina, var. Muhlenbergii. Violet Family.

Three to eight inches high. *Stems.*—Leafy. *Leaves.*—Heart-shaped, wavy-toothed. *Flowers.*—Pale violet.

This is the commonest blue species of the leafy-stemmed violets. It is found in wet, shady places from May till July.

BLUETS. QUAKER LADIES.

Houstonia cærulea. Madder Family.

Stem.—Erect, three to five inches high. *Leaves.*—Very small, opposite. *Flowers.*—Small, delicate blue, lilac, or nearly white, with a yellowish eye. *Calyx.*—Four-lobed. *Corolla.*—Salver-shaped, four-lobed, corolla-tube long and slender. *Stamens.*—Four. *Pistil.*—One, with two stigmas.

No one who has been in the Berkshire Hills during the month of May can forget the loveliness of the bluets. The roadsides, meadows, and even the lawns, are thickly carpeted with the dainty enamel-like blossoms which are always pretty, but which seem to flourish with especial vigor and in great profusion in this lovely region. Less plentiful, perhaps, but still common is the little plant in grassy places far south and west, blossoming in early spring.

The flowers are among those which botanists term "dimorphous." This word signifies *occurring in two forms*, and refers to the stamens and pistils, which vary in size, some flowers having a tall pistil and short stamens, others tall stamens and a short

PLATE LXXXV

BLUETS.—*H. cærulea.*

233

pistil. Darwin has proved, not only that one of these flowers can seldom fully fertilize itself, but that usually the blossoms with tall pistils must be fertilized with pollen from the tall sta-` mens, and that the short pistils are only acted upon by the short stamens. With a good magnifier and a needle these two forms can easily be studied. This is one of the many interesting safe-guards against close-fertilization.

BLUEBELLS. VIRGINIAN COWSLIP. LUNGWORT.

Mertensia Virginica. Borage Family.

One to two feet high. *Stem.*—Smooth, pale, erect. *Leaves.*—Oblong, veiny. *Flowers.*—Blue, pinkish in bud, in raceme-like clusters which are rolled up from the end and straighten as the flowers expand. *Calyx.*—Five-cleft. *Corolla.*—Trumpet-shaped, one inch long, spreading. *Stamens.*—Five. *Pistil.*—One.

These very lovely blossoms are found in moist places during April and May in parts of New York as well as south and west-ward. The English naturalist, Mr. Alfred Wallace, seeing them, for the first time, in the vicinity of Cincinnati, writes in the *Fortnightly Review :* " In a damp river-bottom, the exquisite blue *Mertensia Virginica* was found. It is called here the ' Vir-ginian cowslip,' its drooping porcelain-blue bells being somewhat of the size and form of those of the true cowslip.''

BLUE-EYED MARY. INNOCENCE.

Collinsia verna. Figwort Family.

Six to twenty inches high. *Stems.*—Branching, slender. *Leaves.*—Op-posite, the lower oval, the upper ovate—lance-shaped, clasping by the heart-shaped base. *Flowers.*—Blue and white, long-stalked, appearing whorled in the axils of the upper leaves. *Calyx.*—Deeply five-cleft. *Corolla.*—Deeply two-lipped, the upper lip two-cleft, the lower three-cleft. *Stamens.*—Four. *Pistil.*—One.

Unfortunately these dainty flowers are not found farther east than Western New York. From there they spread south and westward, abounding so plentifully in the vicinity of Cin-cinnati that the moist meadows are blue with their blossoms in spring or early summer.

FORGET-ME-NOT.

Mysotis laxa. Borage Family.

Stems.—Slender. *Leaves.*—Alternate, lance-oblong. *Flowers.*—Blue, small, growing in a raceme. *Calyx.*—Five-lobed. *Corolla.*—Salver-shaped, five-toothed. *Stamens.*—Five. *Pistil.*—One.

Along the banks of the stream, and in low wet places, we may look for these exquisite little flowers. This plant is smaller and less luxuriant than the European species, blossoming in early summer.

WILD PHLOX.

Phlox divaricata. Polemonium Family.

Nine to eighteen inches high. *Stems.*—Spreading or ascending. *Leaves.* —Opposite, oblong or lance-oblong. *Flowers.*—Pale lilac-purple, in a loose, spreading cluster. *Calyx.*—With five slender teeth. *Corolla.*—With a five-parted border, salver-shaped, with a long tube. *Stamens.*—Five, unequally inserted in the tube of the corolla. *Pistil.*—One, with a three-lobed style.

We may search for these graceful, delicately tinted flowers in the rocky woods of April and May.

Nearly allied to them is the wild sweet William, *P. maculata,* the pink-purple blossoms of which are found along the streams and in the rich woods of somewhat southern localities.

The beautiful moss pink, *P. subulata,* is also a member of this genus. This little evergreen heath-like plant clothes the dry hill-sides with a glowing mantle of purple-pink every spring, Southern New York being probably its most northerly range in our Eastern States. Great masses of these flowers may be seen covering the rocks in the Central Park in May.

The pink or whitish clusters of *P. glaberrima* are found in the open woods and prairies somewhat westward in July.

ROBIN'S PLANTAIN. BLUE SPRING-DAISY.

Erigeron bellidifolius. Composite Family (p. 13).

Stem.—Simple, hairy, producing offsets from the base. *Root-leaves.* —Somewhat rounded or wedge-shaped. *Stem-leaves.*—Somewhat oblong, lance-shaped, partly clasping. *Flower-heads.*—Rather large, on slender flower-stalks, composed of both strap-shaped and tubular flowers, the former (ray-flowers) bluish-purple, the latter (disk-flowers) yellow.

This is one of the earliest members of the Composite family to make its appearance, that great tribe being usually associated with the late summer months. The flower might easily be taken for a purple aster which had mistaken the season, or for a blue daisy, as one of its common names suggests. *E. Philadelphicus* is a later very similar species with smaller flower-heads.

ONE-FLOWERED CANCER-ROOT.

Aphyllon uniflorum. Broom-Rape Family.

Scape.—Slender, fleshy, three to five inches high, one-flowered. *Leaves.* —None. *Flower.*—Pale purple, solitary, one inch long, with a delicate fragrance. *Calyx.*—Five-cleft. *Corolla.*—Somewhat two-lipped, with two yellow bearded folds in the throat. *Stamens.*—Four. *Pistil.*—One.

In April or May the odd pretty flower of the parasitic one-flowered cancer-root is found in the damp woodlands.

VIOLET WOOD SORREL.

Oxalis violacea. Geranium Family.

Scape.—Five to nine inches high, several-flowered. *Leaves.*—Divided into three clover-like leaflets. *Flowers.*—Violet-colored, clustered on the scape. *Calyx.*—Of five sepals. *Corolla.*—Of five petals. *Stamens.*—Ten. *Pistil.*—One, with five styles.

This little plant is found in somewhat open or rocky woods, its lovely delicate flower - clusters appearing in May or June. This species is more common southward, while the pink-veined wood sorrel (Pl. XVII.) abounds in the cool woods of the North.

PITCHER PLANT. SIDE-SADDLE FLOWER. HUNTSMAN'S CUP.

Sarracenia purpurea. Pitcher-plant Family.

Scape.—Naked, one-flowered, about one foot high. *Leaves.*—Pitcher-shaped, broadly winged, hooded. *Flower.*—Red-purple, large, nodding. *Calyx.*—Of five colored sepals, with three bractlets at the base. *Corolla.*— Of five fiddle-shaped petals which are arched over the greenish-yellow style. *Stamens.*—Numerous. *Pistil.*—One, with a short style which expands at the summit into a petal-like umbrella-shaped body, with five small hooked stigmas.

The large nodding flower of the pitcher-plant may be found during June in the peat-bogs of New England as well as farther south and west. It is less familiar to most people than the

PLATE LXXXVI

ROBIN'S PLANTAIN.—*E. bellidifolius.*

curious pitcher-like leaves, which are usually partially filled with water and drowned insects ; part of their inner surface being covered with a sugary exudation, below which, for a space, they are highly polished, while on the lower portion grow the stiff bristles which point downward. Insects attracted by the sugary secretion find themselves prisoners, as they can seldom fight their way through the opposing bristles, neither can they usually escape by such a perpendicular flight as would be necessary from the form of the cavity. The decomposed bodies of these unfortunates are supposed to contribute to the nourishment of the plant, as it is hardly probable that this elaborate contrivance answers no special purpose.

WILD GERANIUM. WILD CRANESBILL.

Geranium maculatum. Geranium Family.

Stem.—Erect, hairy. *Leaves.*—About five-parted, the divisions lobed and cut. *Flowers.*—Pale pink-purple, rather large. *Calyx.*—Of five sepals. *Corolla.*—Of five petals. *Stamens.*—Ten. *Pistil.*—With five styles, which split apart at maturity so elastically as to discharge the seeds to some distance.

In spring and early summer the open woods and shaded roadsides are abundantly brightened with these graceful flowers. They are of peculiar interest because of their close kinship with the species, *G. pratense*, which first attracted the attention of the German scholar, Sprengel, to the close relations existing between flowers and insects. The beak-like appearance of its fruit give the plant both its popular and scientific names, for *geranium* is from the Greek for crane. The specific title, *maculatum*, refers to the somewhat blotched appearance of the older leaves.

GILL-OVER-THE-GROUND. GROUND IVY.

Nepeta Glechoma. Mint Family (p. 16).

Stems. — Creeping and trailing. *Leaves.* — Small and kidney - shaped. *Flowers.*—Bluish-purple, loosely clustered in the axils of the leaves. *Calyx.* —Five-toothed. *Corolla.*—Two-lipped, the upper lip erect and two-cleft, the lower spreading and three-cleft. *Stamens.*—Four. *Pistil.*—One, two-lobed at the apex.

As the pleasant aroma of its leaves suggest, this little plant is closely allied to the catnip. Its common title of Gill-over-the

PLATE LXXXVII

WILD GERANIUM.—*G. maculatum.*

ground, appeals to one who is sufficiently without interest in pasture-land (for it is obnoxious to cattle) to appreciate the pleasant fashion in which this little immigrant from Europe has made itself at home here, brightening the earth with such a generous profusion of blossoms every May. But it is somewhat of a disappointment to learn that this name is derived from the French *guiller*, and refers to its former use in the fermentation of beer. Oddly enough the name of alehoof, which the plant has borne in England and which naturally has been supposed to refer to this same custom, is said by a competent authority (Professor Earle, of Oxford) to have no connection with it, but to signify *another sort of hofe, hofe* being the early English name for the violet, which resembles these flowers in color.

The plant was highly prized formerly as a domestic medicine. Gerarde claims that "boiled in mutton-broth it helpeth weake and akeing backs."

LARKSPUR.

Delphinium. Crowfoot Family.

Six inches to five feet high. *Leaves.*—Divided or cut. *Flowers.*—Blue or purplish, growing in terminal racemes. *Calyx.*—Of five irregular petal-like sepals, the upper one prolonged into a spur. *Corolla.*—Of four irregular petals, the upper pair continued backward in long spurs which are enclosed in the spur of the calyx, the lower pair with short claws. *Stamens.*—Indefinite in number. *Pistils.*—One to five, forming pods in fruit.

In April and May the bright blue clusters of the dwarf larkspur, *D. tricorne*, are noticeable in parts of the country. Unfortunately they are not found east of Western Pennsylvania.

The tall wand-like purplish racemes of the tall larkspur, *D. exaltatum*, are found in July in the rich soil of Pennsylvania, and much farther south and west as well.

WILD LUPINE.

Lupinus perennis. Pulse Family (p. 16).

Stem.—Erect, one to two feet high. *Leaves.*—Divided into seven to eleven leaflets. *Flowers.*—Blue, papilionaceous, showy, in a long raceme. *Pod.*—Broad, hairy.

In June and July the long bright clusters of the wild lupine are very noticeable in many of our sandy fields. Its pea-like blossoms serve to easily identify it. Under date of June 8th, Thoreau writes: "The lupine is now in its glory. . . . It paints a whole hill-side with its blue, making such a field (if not meadow) as Proserpine might have wandered in. Its leaf was made to be covered with dew-drops. I am quite excited by this prospect of blue flowers in clumps, with narrow intervals, such a profusion of the heavenly, the Elysian color, as if these were the Elysian fields. . . . That is the value of the lupine. The earth is blued with it."

HAREBELL.

Campanula rotundifolia. Campanula Family.

Stem.—Slender, branching, from five to twelve inches high. *Root-leaves.*—Heart-shaped or ovate, early withering. *Stem-leaves.*—Numerous, long and narrow. *Flowers.*—Bright blue, nodding from hair-like stalks. *Calyx.*—Five-cleft, the lobes awl-shaped. *Corolla.*—Bell-shaped, five-lobed. *Stamens.*—Five. *Pistil.*—One, with three stigmas.

This slender, pretty plant, hung with its tremulous flowers, springs from the rocky cliffs which buttress the river as well as from those which crown the mountain. I have seen the west shore of the Hudson bright with its delicate bloom in June, and the summits of the Catskills tinged with its azure in September. The drooping posture of these flowers protect their pollen from rain or dew. They have come to us from Europe, and are identical, I believe, with the celebrated Scotch bluebells.

BLUE-EYED GRASS.

Sisyrinchium angustifolium. Iris Family.

Four to twelve inches high. *Leaves.*—Narrow and grass-like. *Flowers.*—Blue or purple, with a yellow centre. *Perianth.*—Six-parted, the divisions bristle-pointed. *Stamens.*—Three, united. *Pistil.*—One, with three thread-like stigmas.

> For the sun is no sooner risen with a burning heat,
> But it withereth the grass,
> And the flower thereof falleth,
> And the grace of the fashion of it perisheth.

So reads the passage in the Epistle of James, which seems so graphically to describe the brief life of this little flower, that we might almost believe the Apostle had had it in mind, were it to be found in the East.

The blue-eyed grass belongs to the same family as the showy fleur-de-lis and blossoms during the summer, being especially plentiful in moist meadows. It is sometimes called "eye-bright," which name belongs by rights to *Euphrasia officinalis*.

VENUS'S LOOKING-GLASS.

Specularia perfoliata. Campanula Family.

Stem.—Somewhat hairy, three to twenty inches high. *Leaves.*—Toothed, rounded, clasping by the heart-shaped base. *Flowers.*—Blue. *Calyx.*—Three, four, or five-lobed. *Corolla.*—Wheel-shaped, five-lobed. *Stamens.*—Five. *Pistil.*—One, with three stigmas.

We borrow from Mr. Burroughs's "Bunch of Herbs" a description of this little plant, which blossoms from May till August. "A pretty and curious little weed, sometimes found growing in the edge of the garden, is the clasping specularia, a relative of the harebell and of the European Venus's looking-glass. Its leaves are shell-shaped, and clasp the stalk so as to form little shallow cups. In the bottom of each cup three buds appear that never expand into flowers, but when the top of the stalk is reached, one and sometimes two buds open a large, delicate purple-blue corolla. All the first-born of this plant are still-born as it were ; only the latest, which spring from its summit, attain to perfect bloom."

SKULL-CAP.

Scutellaria. Mint Family (p. 16).

Stem.—Square, usually one or two feet high. *Leaves.*—Opposite, oblong, lance-shaped or linear. *Flowers.*—Blue. *Calyx.*—Two-lipped, the upper lip with a small, helmet-like appendage which at once identifies this genus. *Corolla.*—Two-lipped, the upper lip arched, the lateral lobes mostly connected with the upper lip, the lower lip spreading and notched at the apex. *Stamens.*—Four, in pairs. *Pistil.*—One, with a two-lobed style.

The prettiest and most striking of this genus is the larger skull-cap, *S. integrifolia*, whose bright blue flowers are about one inch long, growing in terminal racemes. In June and July they may be found among the long grass of the roadsides and

PLATE LXXXVIII

BLUE-EYED GRASS.—*S. angustifolium.*

243

meadows. They are easily identified by the curious little appendage on the upper part of the calyx, which gives to this genus its common name.

Perhaps the best-known member of the group is the mad-dog skull-cap, *S. lateriflora*, which delights in wet places, bearing small, inconspicuous flowers in one-sided racemes. This plant is quite smooth, while that of *S. integrifolia* is rather downy. It was formerly believed to be a sure cure for hydrophobia.

S. galericulata is usually found somewhat northward. Its flowers are much larger than those of *S. lateriflora*, but smaller than those of *S. integrifolia*. They grow singly from the axils of the upper leaves.

FLEUR-DE-LIS. LARGER BLUE FLAG.

Iris versicolor. Iris Family.

Stem.—Stout, angled on one side, leafy, one to three feet high. *Leaves.*—Flat and sword-shaped, with their inner surfaces coherent for about half of their length. *Flowers.*—Large and showy, violet-blue, variegated with green, yellow, or white; purple-veined. *Perianth.*—Six-cleft, the three outer divisions recurved, the three inner smaller and erect. *Stamens.*—Three, covered by the three overarching, petal-like divisions of the style. *Pistil.*—One, with its style cleft into three petal-like divisions, each of which bears its stigma on its inner surface.

> Born in the purple, born to joy and pleasance,
> Thou dost not toil nor spin,
> But makest glad and radiant with thy presence
> The meadow and the lin.*

In both form and color this is one of the most regal of our wild flowers, and it is easy to understand why the fleur-de-lis was chosen as the emblem of a royal house, although the especial flower which Louis VII. of France selected as his badge was probably white.

It will surprise most of us to learn that the common name which we have borrowed from the French does not signify "flower-of-the-lily," as it would if literally translated, but "flower of Louis," *lis* being a corruption of the name of the king who first adopted it as his badge.

For the botanist the blue-flag possesses special interest. It

* Longfellow.

PLATE LXXXIX

FLEUR-DE-LIS.—*I. versicolor.*

is a conspicuous example of a flower which has guarded itself against self-fertilization, and which is beautifully calculated to secure the opposite result. The position of the stamens is such that their pollen could not easily reach the stigmas of the same flower, for these are borne on the inner surface of the petal-like, overarching styles. There is no prospect here of any seed being set unless the pollen of another flower is secured. Now what are the chances in favor of this? They are many : In the first place the blossom is unusually large and showy, from its size and shape alone almost certain to arrest the attention of the passing bee ; next, the color is not only conspicuous, but it is also one which has been found to be especially attractive to bees ; blue and purple flowers being particularly sought by these insects. When the bee reaches the flower he alights on the only convenient landing-place, one of the recurved sepals ; following the deep purple veins which experience has taught him lead to the hidden nectar, he thrusts his head below the anther, brushing off its pollen, which he carries to another flower.

The rootstocks of the Florentine species of iris yield the familiar " orris-root."

The family name is from the Greek for *rainbow*, on account of the rich and varied hues of its different members.

The plant abounds in wet meadows, the blossoms appearing in June.

AMERICAN BROOKLIME.

Veronica Americana.   Figwort Family.

Stem.—Smooth, reclining at base, then erect, eight to fifteen inches high. *Leaves.*—Mostly opposite, oblong, toothed.   *Flowers.*—Blue, clustered in the axils of the leaves.   *Calyx.*—Four-parted.   *Corolla.*—Wheel-shaped, four-parted.   *Stamens.*—Two.   *Pistil.*—One.

Perhaps the prettiest of the blue *Veronicas* is the American brooklime. Its clustered flowers make bright patches in moist ground which might, at a little distance, be mistaken for beds of forget-me-nots. It blossoms from June till August, and is almost as common in wet ditches and meadows as its sister, the common speedwell, is in dry and open places. Some of the

PLATE XC

AMERICAN BROOKLIME.—*V. Americana.*

247

members of this genus were once believed to possess great medicinal virtues, and won for themselves in Europe the laudatory names of Honor and Praise.

COMMON SPEEDWELL.

Veronica officinalis. Figwort Family.

The little speedwell's darling blue

is noticeable during June and July, when clusters of these tiny flowers brighten many a waste spot along the sunny roadsides. This is a hairy little plant, with a stem which lies upon the ground and takes root, thus spreading itself quickly over the country.

ARETHUSA.

Arethusa bulbosa. Orchis Family (p. 17).

Scape.—Sheathed, from a globular bulb, usually one-flowered. *Leaf.*—"Solitary, linear, nerved, hidden in the sheaths of the scape, protruding after flowering." (Gray.) *Flower.*—Rose-purple, large, with a bearded lip.

In some localities this beautiful flower is very plentiful. Every June will find certain New England marshes tinged with its rose-purple blossoms, while in other near and promising bogs it may be sought vainly for years. At least it may be hoped for in wet places as far south as North Carolina, its most favorite haunt being perhaps a cranberry-swamp. Concerning it, Mr. Burroughs writes: "Arethusa was one of the nymphs who attended Diana, and was by that goddess turned into a fountain, that she might escape the god of the river Alpheus, who became desperately in love with her on seeing her at her bath. Our Arethusa is one of the prettiest of the orchids, and has been pursued through many a marsh and quaking - bog by her lovers. She is a bright pink-purple flower an inch or more long, with the odor of sweet violets. The sepals and petals rise up and arch over the column, which we may call the heart of the flower, as if shielding it. In Plymouth County, Mass., where the Arethusa seems common, I have heard it called Indian pink."

PURPLE FRINGED ORCHISES.

Orchis Family (p. 17).

Habenaria fimbriata.

Leaves.—Oval or oblong ; the upper, few, passing into lance-shaped bracts. *Flowers.*—Purple, rather large ; with a fan-shaped, three-parted lip, its divisions fringed ; with a long curving spur ; growing in a spike.

Habenaria psycodes.

Leaves.—Oblong or lance-shaped, the upper passing into linear bracts. *Flowers.*—Purple, fragrant, resembling those of *H. fimbriata,* but much smaller, with a less fringed lip ; growing in a spike.

We should search the wet meadows in early June if we wish to be surely in time for the larger of the purple fringed orchises, for *H. fimbriata* somewhat antedates *H. psycodes,* which is the commoner species of the two and appears in July. Under date of June 9th, Thoreau writes : " Find the great fringed-orchis out apparently two or three days, two are almost fully out, two or three only budded ; a large spike of peculiarly delicate, pale-purple flowers growing in the luxuriant and shady swamp, amid hellebores, ferns, golden senecio, etc. . . . The village belle never sees this more delicate belle of the swamp. . . . A beauty reared in the shade of a convent, who has never strayed beyond the convent-bell. Only the skunk or owl, or other inhabitant of the swamp, beholds it."

AMERICAN PENNYROYAL.

Hedeoma pulegioides. Mint Family (p. 16).

Stem.—Square, low, erect, branching, *Leaves.*—Opposite, aromatic, small. *Flowers.*—Purplish, small, whorled in the axils of the leaves. *Calyx.*—Two-lipped, upper lip three-toothed, the lower two-cleft. *Corolla.*—Two-lipped, upper erect, notched at apex, the lower spreading and three-cleft. *Fertile stamens.*—Two. *Pistil.*—One, with a two-lobed style.

This well-known, strong-scented little plant is found throughout the greater part of the country, blossoming in midsummer.

249

Its taste and odor nearly resemble that of the true pennyroyal, *Mentha pulegium*, of Europe.

MONKEY-FLOWER.

Mimulus ringens. Figwort Family.

Stem.—Square, one to two feet high. *Leaves.*—Opposite, oblong or lance-shaped. *Flowers.*—Pale violet-purple, rarely white, growing singly from the axils of the leaves. *Calyx.*—Five-angled, five-toothed, the upper tooth largest. *Corolla.*—Tubular, two-lipped, the upper lip erect or spreading, two-lobed, the lower spreading and three-lobed, the throat closed. *Stamens.*—Four. *Pistil.*—One, with a two-lobed stigma.

From late July onward the monkey-flowers tinge the wet fields and border the streams and ponds ; not growing in the water like the pickerel-weed, but seeking a hummock in the swamp, or a safe foothold on the brook's edge, where they can absorb the moisture requisite to their vigorous growth.

The name is a diminutive of *mimus*—a buffoon, and refers to the somewhat grinning blossom. The plant is a common one throughout the eastern part of the country.

COMMON MOTHERWORT.

Leonurus cardiaca. Mint Family (p. 16).

Stem.—Tall and upright. *Leaves.*—Opposite, the lower rounded and lobed, the floral wedge-shaped at base and three-cleft. *Flowers.*—Pale purple, in close whorls in the axils of the leaves. *Calyx.*—"With five nearly equal teeth, which are awl-shaped, and when old rather spiny, pointed, and spreading." (Gray.) *Corolla.*—Two-lipped, the upper lip somewhat arched and bearded, the lower three-lobed and spreading. *Stamens.*—Four, in pairs. *Pistil.*—One, with a two-lobed style.

The tall erect stems, opposite leaves, and regular whorls of closely clustered pale purple flowers help us to easily identify the motherwort, if identification be needed, for it seems as though such old-fashioned, time-honored plants as catnip, tansy, and motherwort, which cling so persistently to the skirts of the old homestead in whose domestic economy they once played so important a part, should be familiar to us all.

PLATE XCI

MONKEY-FLOWER.—*M. ringens.*

CORN COCKLE.

Lychnis Githago.   Pink Family.

About two feet high. *Leaves.*—Opposite, long and narrow, pale green, with silky hairs. *Flowers.*—Rose-purple, large, long-stalked. *Calyx-lobes.* —Five, long and slender, exceeding the petals. *Corolla.*—Of five broad petals. *Stamens.*—Ten. *Pistil.*—One, with five styles.

In many countries some of the most beautiful and noticeable flowers are commonly found in grain-fields. England's scarlet poppies flood her farm-lands with glorious color in early summer; while the bluets lighten the corn-fields of France. Our grain-fields seem to have no native flower peculiar to them; but often we find a trespasser of foreign descent hiding among the wheat or straying to the roadsides in early summer, whose deep-tinted blossoms secure an instant welcome from the flower-lover if not from the farmer. "What hurte it doeth among corne! the spoyle unto bread, as well in colour, taste, and unwholesomeness, is better known than desired," wrote Gerarde. The large dark seeds fill the ground wheat with black specks, and might be injurious if existing in any great quantity. Its former generic name was *Agrostemma*, signifying *crown of the fields.* Its present one of *Lychnis*, signifies *a light* or *lamp.*

BLUE VERVAIN.   SIMPLER'S JOY.

Verbena hastata.   Vervain Family.

Four to six feet high. *Leaves.*—Opposite, somewhat lance-shaped, the lower often lobed and sometimes halberd-shaped at base. *Flowers.*—Purple, small, in slender erect spikes. *Calyx.*—Five-toothed. *Corolla.*—Tubular, somewhat unequally five-cleft. *Stamens.*—Two, in pairs. *Pistil.*—One.

Along the roadsides in midsummer we notice these slender purple spikes, the appearance of which would be vastly improved if the tiny blossoms would only consent to open simultaneously.

In earlier times the vervain was beset with classic associations. It was claimed as the plant which Virgil and other poets mention as being used for altar-decorations and for the garlands of sacrificial beasts. It was believed to be the *herba sacra* of the ancients, until it was understood that the generic title *Verbena* was a word which was applied to branches of any de-

PLATE XCII

BLUE VERVAIN.—*V. hastata.*

253

scription which were used in religious rites. It certainly seems, however, to have been applied to some especial plant in the time of Pliny, for he writes that no plant was more honored among the Romans than the sacred *Verbena.* In more modern times as well the vervain has been regarded as an "herb of grace," and has been gathered with various ceremonies and with the invocation of a blessing, which began as follows:

> Hallowed be thou, Vervain,
> As thou growest on the ground,
> For in the Mount of Calvary
> There thou was first found.

It was then supposed to be endued with especial virtue, and was worn on the person to avert disaster.

The time-honored title of Simpler's joy arose from the remuneration which this popular plant brought to the "Simplers"—as the gatherers of medicinal herbs were entitled.

BEARD-TONGUE.

Pentstemon pubescens. Figwort Family.

Stem.—One or two feet high, clammy above. *Leaves.*—Opposite, ob-. long to lance-shaped. *Flowers.*—Dull purple or partly whitish, showy, in a slender open cluster. *Calyx.*—Five-parted. *Corolla.*—Tubular, slightly dilated, the throat nearly closed by a bearded palate; two-lipped, the upper lip two-lobed, the lower three-cleft. *Stamens.*—Four, one densely bearded sterile filament besides. *Pistil.*—One.

These handsome, showy flowers are found in summer in dry or rocky places. They are especially plentiful somewhat southward.

The white beard-tongue of more western localities is *P. digitalis.* This is a very effective plant, which sometimes reaches a height of five feet, having large inflated white flowers.

SELF-HEAL. HEAL-ALL.

Brunella vulgaris. Mint• Family (p. 16).

Stems.—Low. *Leaves.*—Opposite, oblong. *Flowers.*—Bluish-purple, in a spike or head. *Calyx.*—Two-lipped, upper lip with three short teeth, the lower two-cleft *Corolla.*—Two-lipped, the upper lip arched, entire, the lower spreading, three-cleft. *Stamens.*—Four. *Pistil.*—One, two-lobed at the apex.

Throughout the length and breadth of the country, from June until September, the short, close spikes of the self-heal can

PLATE XCIII

SELF-HEAL.—*B. vulgaris.*

be found along the roadsides. The botanical name, *Brunella*, is a corruption from *Prunella*, which is taken from the German for quinsy, for which this plant was considered a certain cure. It was also used in England as an application to the wounds received by rustic laborers, as its common names, carpenter's herb, hook-heal, and sicklewort, imply. That the French had a similar practice is proved by an old proverb of theirs to the effect that " No one wants a surgeon who keeps *Prunelle*."

WILD BERGAMOT.

Monarda fistulosa. Mint Family (p. 16).

Two to five feet high. *Leaves.*--Opposite, fragrant, toothed. *Flowers.*—Purple or purplish, dotted, growing in a solitary, terminal head. *Calyx.*—Tubular, elongated, five-toothed. *Corolla.*—Elongated, two-lipped. *Stamens.*—Two, elongated. *Pistil.*—One, with style two-lobed at apex.

Although the wild bergamot is occasionally found in our eastern woods, it is far more abundant westward, where it is found in rocky places in summer. This is a near relative of the bee balm (Pl. LXXXII.), which it closely resembles in its manner of growth.

DAY-FLOWER.

Commelina Virginica. Spiderwort Family.

Stem.—Slender, branching. *Leaves.*—Lance-shaped to linear, the floral ones heart-shaped and clasping, folding so as to enclose the flowers. *Flowers.*—Blue. *Calyx.*—Of three unequal somewhat colored sepals, the two lateral ones partly united. *Corolla.*—Of three petals, two large, rounded, pale blue, one small, whitish, and inconspicuous. *Stamens.*—Six, unequal in size, three small and sterile, with yellow cross-shaped anthers, three fertile, one of which is bent inward. *Pistil.*—One.

The odd day-flower is so named because its delicate blossoms only expand for a single morning. At the first glance there seem to be but two petals which are large, rounded, and of a delicate shade of blue. A closer examination, however, discovers still another, so inconspicuous in form and color as to escape the notice of the casual observer. This inequality recalls the quaint tradition as to the origin of the plant's generic name. There were three brothers Commelin, natives of Holland. Two of them were botanists of repute, while the tastes of the third had a less marked botanical tendency. The genus was dedicated to the trio : the two large bright petals commemorating the brother botanists, while the small and unpretentious one perpetuates the memory of him who was so unwise as to take little or no interest in so noble a science. These flowers appear throughout the summer in cool woods and on moist banks.

BLUE LINARIA. BLUE TOADFLAX.
Linaria Canadensis. Figwort Family.

Stems.—Slender, six to thirty inches high. *Leaves.*—Linear. *Flowers.* —Pale blue or purple, small, in a long terminal raceme. *Calyx.*—Five-parted. *Corolla.*—Two-lipped, with a slender spur, closed in the throat. *Stamens.*—Four. *Pistil.*—One.

The slender spikes of the blue linaria flank the sandy road-sides nearly all summer, and even in November we find a few delicate blossoms still left upon the elongated stems. These flowers have a certain spirituality which is lacking in their handsome, self-assertive relation, butter-and-eggs.

SPIDERWORT.
Tradescantia Virginica. Spiderwort Family.

Stems.—Mucilaginous, leafy, mostly upright. *Leaves.*—Linear, keeled. *Flowers.*—Blue, clustered, with floral leaves as in the day-flower. *Calyx.*— Of three sepals. *Corolla.*—Of three petals. *Stamens.*—Six, with bearded filaments. *Pistil.*—One.

The flowers of the spiderwort, like those of the day-flower, to which they are nearly allied, are very perishable, lasting only a few hours. They are found throughout the summer, somewhat south and westward. The genus is named in honor of Tradescant, gardener to Charles I. of England.

PICKEREL-WEED.
Pontedaria cordata. Pickerel-weed Family.

Stem.—Stout, usually one-leaved. *Leaves.*—Arrow or heart-shaped. *Flowers.*—Blue, fading quickly, with an unpleasant odor, growing in a dense spike. *Perianth.*—Two-lipped, the upper lip three-lobed and marked with a double greenish-yellow spot, the lower of three spreading divisions. *Stamens.*—Six, three long and protruding, the three others, which are often im-perfect, very short and inserted lower down. *Pistil.*—One.

The pickerel-weed grows in such shallow water as the pick-erel seek, or else in moist, wet places along the shores of streams and rivers. We can look for the blue, closely spiked flowers from late July until some time in September. They are often found near the delicate arrow-head.

BLUEWEED. VIPER'S BUGLOSS.

Echium vulgare. Borage Family.

Stem.—Rough, bristly, erect, about two feet high. *Leaves.*—Alternate, lance-shaped, set close to the stem. *Flowers.*—Bright blue, spiked on one side of the branches, which are at first rolled up from the end, straightening as the blossoms expand. *Calyx.*—Five-parted. *Corolla.*—Of five somewhat unequal, spreading lobes. *Stamens.*—Five, protruding, red. *Pistil.*—One.

When the blueweed first came to us from across the sea it secured a foothold in Virginia. Since then it has gradually worked its way northward, lining the Hudson's shores, overrunning many of the dry fields in its vicinity, and making itself at home in parts of New England. We should be obliged to rank it among the " pestiferous " weeds were it not that, as a rule, it only seeks to monopolize land which is not good for very much else. The pinkish buds and bright blue blossoms with their red protruding stamens make a valuable addition, from the æsthetic point of view, to the bunch of midsummer field-flowers in which hitherto the various shades of red and yellow have predominated.

NIGHTSHADE.

Solanum Dulcamara. Nightshade Family.

Stem.—Usually somewhat climbing or twining. *Leaves.*—Heart-shaped, the upper halberd-shaped or with ear-like lobes or leaflets at the base. *Flowers.*—Purple, in small clusters. *Calyx.*—Five-parted. *Corolla.*—Five-parted, wheel-shaped. *Stamens.*—Five, yellow, protruding. *Pistil.*—One. *Fruit.*—A red berry.

The purple flowers, which at once betray their kinship with the potato plant, and, in late summer, the bright red berries of the nightshade, cluster about the fences and clamber over the moist banks which line the highway. This plant, which was imported from Europe, usually indicates the presence of civilization. It is not poisonous to the touch, as is often supposed, and it is doubtful if the berries have the baneful power attributed to them.

PLATE XCIV

BLUEWEED.—*E. vulgare.*

259

Thoreau writes regarding them: "The Solanum Dulcamara berries are another kind which grow in drooping clusters. I do not know any clusters more graceful and beautiful than these drooping cymes of scented or translucent, cherry-colored elliptical berries. . . . They hang more gracefully over the river's brim than any pendant in a lady's ear. Yet they are considered poisonous; not to look at surely. . . . But why should they not be poisonous? Would it not be bad taste to eat these berries which are ready to feed another sense?"

GREAT LOBELIA.

Lobelia syphilitica. Lobelia Family.

Stem.—Leafy, somewhat hairy, one to three feet high. *Leaves.*—Alternate, ovate to lance-shaped, thin, irregularly toothed. *Flowers.*—Rather large, light blue, spiked. *Calyx.*—Five-cleft, with a short tube. *Corolla.*—Somewhat two-lipped, the upper lip of two rather erect lobes, the lower spreading and three-cleft. *Pistil.*—One, with a fringed stigma.

The great lobelia is a striking plant which grows in low ground, flowering in midsummer. In some places it is called "High-Belia," a pun which is supposed to reflect upon the less tall and conspicuous species, such as the Indian tobacco, *L. inflata*, which are found flowering at the same season.

If one of its blossoms is examined, the pistil is seen to be enclosed by the united stamens in such a fashion as to secure self-fertilization, one would suppose. But it is hardly probable that a flower as noticeable as this, and wearing a color as popular as blue, should have adorned itself so lavishly to no purpose. Consequently we are led to inquire more closely into its domestic arrangements. Our curiosity is rewarded by the discovery that the lobes of the stigma are so tightly pressed together that they can at first receive no pollen upon their sensitive surfaces. We also find that the anthers open only by a pore at their tips, and when irritated by the jar of a visiting bee, discharge their pollen upon its body through these outlets. This being accomplished the fringed stigma pushes forward, brushing aside whatever pollen may have fallen within the tube. When it finally

PLATE XCV

GREAT LOBELIA.—*L. syphilitica.*

.261

projects beyond the anthers, it opens, and is ready to receive its pollen from the next insect-visitor.

The genus is named after an early Flemish herbalist, de l'Obel.

INDIAN TOBACCO.

Lobelia inflata. Lobelia Family.

One to two feet high. *Stem.*—Branching from the root. *Leaves.*—Alternate, ovate, or oblong, somewhat toothed. *Flowers.*—Blue or purple, small, growing in a loose raceme, resembling in structure those of the great lobelia. *Pod.*—Much inflated.

During the summer we note in the dry, open fields the blue racemes of the Indian tobacco, and in the later year the inflated pods which give it its specific name. The plant is said to be poisonous if taken internally, and yields a " quack-medicine " of some notoriety. The Indians smoked its dried leaves, which impart to the tongue a peculiar tobacco-like sensation.

There are other species of lobelia which may be distinguished by their narrower leaves and uninflated pods, and by their choice of moist localities.

HOG PEA-NUT.

Amphicarpæa monoica. Pulse Family (p. 16).

Stem.—Climbing and twining over plants and shrubs. *Leaves.*—Divided into three somewhat four-sided leaflets. *Flowers.*—Papilionaceous, pale lilac, or purplish, in nodding racemes. *Pod.*—One inch long.

Along the shadowy lanes which wind through the woods the climbing members of the Pulse family are very abundant. During the late summer and autumn the lonely wayside is skirted by

> Vines, with clust'ring bunches growing ;
> Plants, with goodly burden bowing.

And in and out among this luxuriant growth twist the slender stems of the ill-named hog pea-nut, its delicate lilac blossoms nodding from the coarse stalks of the golden-rods and iron-weeds or blending with the purple asters.

This plant bears flowers of two kinds : the upper ones are perfect, but apparently useless, as they seldom ripen fruit ; while the lower or subterranean ones are without petals or attractiveness of appearance, but yield eventually at least one large ripe seed.

PLATE XCVI

INDIAN TOBACCO.—*L. inflata.*

263

BEACH PEA.

Lathyrus maritimus. Pulse Family (p. 16).

About one foot high, or more. *Stem.*—Stout. *Leaves.*—Divided into from three to five pairs of thick oblong leaflets. *Flowers.*—Papilionaceous, large, purple, clustered.

The deep-hued flowers of this stout plant are commonly found along the sand-hills of the seashore, and also on the shores of the Great Lakes, blooming in early summer. Both flowers and leaves are at once recognized as belonging to the Pulse family.

Strophostyles angulosa. Pulse Family (p. 16).

Stems.—Branched, one to six feet long, prostrate or climbing. *Leaves.*— Divided into three leaflets, which are more or less prominently lobed toward the base, the terminal two-lobed ; or some or all without lobes. *Flowers.*— Purplish or greenish, on long flower-stalks. *Pod.*—Linear, straight, or nearly so.

This somewhat inconspicuous plant is found back of the sand-hills along the coast, often in the neighborhood of the beach pea, and climbing over river-banks, thickets, and fences as well. It can usually be identified by its oddly lobed leaflets.

BLUE VETCH.

Vicia cracca. Pulse Family (p. 16).

Leaves.—Divided into twenty to twenty-four leaflets, with slender tips. *Flowers.*—Papilionaceous, blue turning purple, growing in close, many-flowered, one-sided spikes.

This is an emigrant from Europe which is found in some of our eastern fields and thickets as far south as New Jersey. It usually climbs more or less by means of the tendril at the tip of its divided leaves, and sometimes forms bright patches of vivid blue over the meadows.

Another member of this genus is *V. sativa*, the common vetch or tare, with purplish or pinkish flowers, growing singly or in pairs from the axils of the leaves, which leaves are divided

PLATE XCVII

BEACH PEA.—*L. maritimus.*

265

into fewer and narrower leaflets than those of the blue vetch. This species also takes possession of cultivated fields as well as of waste places along the roadside.

CHICORY. SUCCORY.

Cichorium Intybus. Composite Family (p. 13).

Stems.—Branching; *Leaves.*—The lower oblong or lance-shaped, partly clasping, sometimes sharply incised, the floral ones minute. *Flower-heads.* —Blue, set close to the stem, composed entirely of strap-shaped flowers ; opening at different times.

> Oh, not in Ladies' gardens,
> My peasant posy !
> Smile thy dear blue eyes,
> Nor only—nearer to the skies—
> In upland pastures, dim and sweet,—
> But by the dusty road
> Where tired feet
> Toil to and fro ;
> Where flaunting Sin
> May see thy heavenly hue,
> Or weary Sorrow look from thee
> Toward a more tender blue !*

This roadside weed blossoms in late summer. It is extensively cultivated in France, where the leaves are blanched and used in a salad which is called "Barbe des Capucins." The roots are roasted and mixed with coffee both there and in England.

Horace mentions its leaves as part of his frugal fare, and Pliny remarks upon the importance of the plant to the Egyptians, who formerly used it in great quantities, and of whose diet it is still a staple article.

BLUE AND PURPLE ASTERS.

Aster. Composite Family (p. 13).

Flower-heads.—Composed of blue or purple ray-flowers, with a centre of yellow disk-flowers.

As about one hundred and twenty different species of aster are native to the United States, and as fifty-four of these are found in Northeastern America, all but a dozen being purple or

* Margaret Deland.

PLATE XCVIII

Single flower.

CHICORY.—*C. Intybus.*

blue (*i.e.*, with purple or blue ray-flowers), and as even botanists find that it requires patient application to distinguish these
many species, only a brief description of the more conspicuous
and common ones is here attempted.

Along the dry roadsides in early August we may look for the
bright blue-purple flowers of *A. patens*. This is a low-growing
species, with rough, narrowly oblong, clasping leaves, and widely
spreading branches, whose slender branchlets are usually terminated by a solitary flower-head.

Probably no member of the group is more striking than the
New England aster, *A. Novæ Angliæ*, whose stout hairy stem
(sometimes eight feet high), numerous lance-shaped leaves, and
large violet-purple or sometimes pinkish flower-heads, are conspicuous in the swamps of late summer.

A. puniceus is another tall swamp-species, with long showy
pale lavender ray-flowers.

One of the most commonly encountered asters is *A. cordifo-
lius*, which is far from being the only heart-leaved species, despite its title. Its many small, pale blue or almost white flower-
heads mass themselves abundantly along the wood-borders and
shaded roadsides.

Perhaps the loveliest of all the tribe is the seaside purple
aster, *A. spectabilis*, a low plant with narrowly oblong leaves
and large bright heads, the violet-purple ray-flowers of which
are nearly an inch long. This grows in sandy soil near the
coast and may be found putting forth its royal, daisy-like blossoms into November.

Great Britain can claim but one native aster, *A. Tripolium*,
or sea-starwort as it is called. Many American species are cultivated in English gardens under the general title of Michaelmas
daisies. The starwort of Italy is *A. amellus*. The Swiss species is *A. Alpinum*.

This beautiful genus, like that of the golden-rod, is one of
the peculiar glories of our country. Every autumn these two
kinds of flowers clothe our roadsides and meadows with so regal
a mantle of purple and gold that we cannot but wonder if the
flowers of any other region combine in such a radiant display.

IRON-WEED.

Vernonia Noveboracensis. Composite Family (p. 13).

Stem.—Leafy, usually tall. *Leaves.*—Alternate, somewhat lance-oblong. *Flower-heads.*—An intense red-purple, loosely clustered, composed entirely of tubular flowers.

. Along the roadsides and low meadows near the coast the iron-weed adds its deep purple hues to the color-pageant of late August. By the uninitiated the plant is often mistaken for an aster, but a moment's inspection will discover that the minute flowers which compose each flower-head are all tubular in shape, and that the ray or strap-shaped blossoms which an aster must have are wanting. These flower-heads are surrounded by an involucre composed of small scales which are tipped with a tiny point and are usually of a purplish color also.

BLUE CURLS. BASTARD PENNYROYAL.

Trichostema dichotomum. Mint Family (p. 16).

Stem.—Rather low, branching, clammy. *Leaves.*—Opposite, narrowly oblong or lance-shaped, glutinous, with a balsamic odor. *Flowers.*—Purple, occasionally pinkish, not usually clustered. *Calyx.*—Five-cleft, two-lipped. *Corolla.*—Five-lobed, the three lower lobes more or less united. *Stamens.*—Four, very long and curved, protruding. *Pistil.*—One, with a two-lobed style.

In the sandy fields of late summer this little plant attracts notice by its many purple flowers. Its corolla soon falls and exposes to view the four little nutlets of the ovary lying within the enlarged calyx like tiny eggs in their nest. Its aromatic odor is very perceptible, and the little glands with which it is covered may be seen with the aid of a magnifier. The generic name, *Trichostema*, signifies *hairy stamens* and alludes to the curved hair-like filaments.

SEA LAVENDER. MARSH ROSEMARY.

Statice Caroliniana. Leadwort Family.

Stems.—Leafless, branching. *Leaves.*—From the root, somewhat oblong, thick. *Flowers.*—Lavender-color or pale purple, tiny, scattered or loosely spiked along one side of the branches. *Calyx.*—Dry, funnel-form. *Corolla.*—Small, with five petals. *Stamens.*—Five. *Pistil.*—One, with five, rarely three, styles.

In August many of the salt marshes are blue with the tiny flowers of the sea lavender. The spray-like appearance of the

269

little plant would seem to account for its name of rosemary, which is derived from the Latin for *sea-spray*, but Dr. Prior states that this name was given it on account of "its usually growing on the sea-coast, and its odor."

Blossoming with the lavender we often find the great rose mallows and the dainty sea pinks. The marsh St. John's-wort as well is frequently a neighbor, and, a little later in the season, the salt marsh fleabane.

BLAZING STAR.

Liatris scariosa. Composite Family (p. 13).

Stem.—Simple, stout, hoary, two to five feet high. *Leaves.*—Alternate, narrowly lance-shaped. *Flower-heads.*—Racemed along the upper part of the stem, composed entirely of tubular flowers of a beautiful shade of rose-purple.

These showy and beautiful flowers lend still another tint to the many-hued salt marshes and glowing inland meadows of the falling year. Gray assigns them to dry localities from New England to Minnesota and southward, while my own experience of them is limited to the New England coast, where their stout leafy stems and bright-hued blossoms are noticeable among the golden-rods and asters of September. The hasty observer sometimes confuses the plant with the iron-weed, but the two flowers are very different in color and in their manner of growth.

COMMON DITTANY.

Cunila Mariana. Mint Family (p. 16).

About one foot high. *Stem.*—Much branched, reddish. *Leaves.*—Opposite, aromatic, dotted, smooth, ovate, rounded or heart-shaped at base, set close to the stem. *Flowers.*—Small, purple, lilac or white, clustered. *Calyx.*—Five-toothed. *Corolla.*—Small, two-lipped, the upper lip erect, usually notched, the lower three-cleft. *Stamens.*—Two, erect, protruding. *Pistil.*—One, with a two-lobed style.

In late August or early September the delicate flowers of the dittany brighten the dry, sterile banks which flank so many of our roadsides. At a season when few plants are flowering save the omnipresent members of the great Composite family these dainty though unpretentious blossoms are especially attractive. The plant has a pleasant fragrance.

PLATE XCIX

BLAZING STAR.—*L. scariosa.*

271

CLOSED GENTIAN.

Gentiana Andrewsii. Gentian Family.

Stem.—One to two feet high, upright, smooth. *Leaves.*—Opposite, narrowly oval or lance-shaped. *Flowers.*—Blue to purple, clustered at the summit of the stem and often in the axils of the leaves. *Calyx.*—Four or five-cleft. *Corolla.*—Closed at the mouth, large, oblong. *Stamens.*—Four or five. *Pistil.*—One, with two stigmas.

Few flowers adapt themselves better to the season than the closed gentian. We look for it in September when the early waning days and frost-suggestive nights prove so discouraging to the greater part of the floral world. Then in somewhat moist, shaded places along the roadside we find this vigorous, autumnal-looking plant, with stout stems, leaves that bronze as the days advance, and deep-tinted flowers firmly closed as though to protect the delicate reproductive organs within from the sharp touches of the late year.

To me the closed gentian usually shows a deep blue or even purple countenance, although like the fringed gentian and so many other flowers its color is lighter in the shade than in the sunlight. But Thoreau claims for it a "transcendent blue," "a splendid blue, light in the shade, turning to purple with age." "Bluer than the bluest sky, they lurk in the moist and shady recesses of the banks," he writes. Mr. Burroughs also finds it "intensely blue."

FIVE-FLOWERED GENTIAN.

Gentiana quinqueflora. Gentian Family.

Stem.—Slender, branching, one or two feet high. *Leaves.*—Opposite, ovate, lance-shaped, partly clasping. *Flowers.*—Pale blue, smaller than those of the closed gentian, in clusters of about five at the summit of stems and branches. *Calyx.*—Four or five-cleft, small. *Corolla.*—Funnel-form, four or five-lobed, its lobes bristle-pointed. *Stamens.*—Four or five. *Pistil.*—One, with two stigmas.

Although the five-flowered gentian is far less frequently encountered than the closed gentian, it is very common in certain localities. Gray assigns it to "moist hills" and "along the mountains to Florida." I have found it growing in great abundance on the Shawangunk Mountains in Orange County, N. Y., where it flowers in September.

PLATE C

CLOSED GENTIAN.—*G. Andrewsii.*

273

FRINGED GENTIAN.

Gentiana crinita. Gentian Family.

Stem.—One to two feet high. *Leaves.*—Opposite, lance-shaped or nar-
rowly oval. *Flowers.*—Blue, large. *Calyx.*—Four-cleft, the lobes unequal.
Corolla.—Funnel-form, with four fringed, spreading lobes. *Stamens.*—Four.
Pistil.—One, with two stigmas.

In late September when we have almost ceased to hope for
new flowers we are in luck if we chance upon this

whose
> —blossom bright with autumn dew

> —sweet and quiet eye
> Looks through its fringes to the sky,
> Blue—blue—as if that sky let fall,
> A flower from its cerulean wall ; *

for the fringed gentian is fickle in its habits, and the fact that we
have located it one season does not mean that we will find it in
the same place the following year ; being a biennial, with seeds
that are easily washed away, it is apt to change its haunts from
time to time. So our search for this plant is always attended
with the charm of uncertainty. Once having ferreted out its
new abiding-place, however, we can satiate ourselves with its
loveliness, which it usually lavishes unstintingly upon the moist
meadows which it has elected to honor.

Thoreau describes its color as "such a dark blue ! surpassing
that of the male bluebird's back ! " My experience has been
that the flowers which grow in the shade are of a clear pure
azure, " Heaven's own blue," as Bryant claims ; while those
which are found in open, sunny meadows may be justly said to
vie with the back of the male bluebird. If the season has been
a mild one we shall perhaps find a few blossoms lingering into
November, but the plant is probably blighted by a severe frost,
although Miss Emily Dickinson's little poem voices another
opinion :

. . . . . .
But just before the snows
There came a purple creature
That ravished all the hill :
And Summer hid her forehead,

And mockery was still.
The frosts were her condition :
The Tyrian would not come
Until the North evoked it,
"Creator ! shall I bloom ? "

* Bryant.

PLATE CI

FRINGED GENTIAN.—*G. crinita.*

VI

MISCELLANEOUS

SKUNK CABBAGE. SWAMP CABBAGE

Symplocarpus fœtidus. Arum Family.

Leaves.—Large, becoming one or two feet long ; heart-shaped, appearing later than the purple-mottled spathe and hidden flowers. *Flowers.*—Small and inconspicuous ; packed on the fleshy spike which is hidden within the spathe.

If we are bold enough to venture into certain swampy places in the leafless woods and brown cheerless meadows of March, we notice that the sharply pointed spathes of the skunk cabbage have already pierced the surface of the earth. Until I chanced upon a passage in Thoreau's Journal under date of October 31st, I had supposed that these "hermits of the bog" were only encouraged to make their appearance by the advent of those first balmy, spring-suggestive days which occasionally occur as early as February. But it seems that many of these young buds had pushed their way upward before the winter set in, for Thoreau counsels those who are afflicted with the melancholy of autumn to go to the swamps, " and see the brave spears of skunk-cabbage buds already advanced toward a new year." " Mortal and human creatures must take a little respite in this fall of the year," he writes. " Their spirits do flag a little. There is a little questioning of destiny, and thinking to go like cowards to where the weary shall be at rest. But not so with the skunk-cabbage. Its withered leaves fall and are transfixed by a rising bud. Winter and death are ignored. The circle of life is complete. Are these false prophets? Is it a lie or a vain boast underneath the skunk-cabbage bud pushing it upward and lifting the dead leaves with it? "

The purplish shell-like leaf, which curls about the tiny flowers which are thus hidden from view, is a rather grewsome-looking

PLATE CII

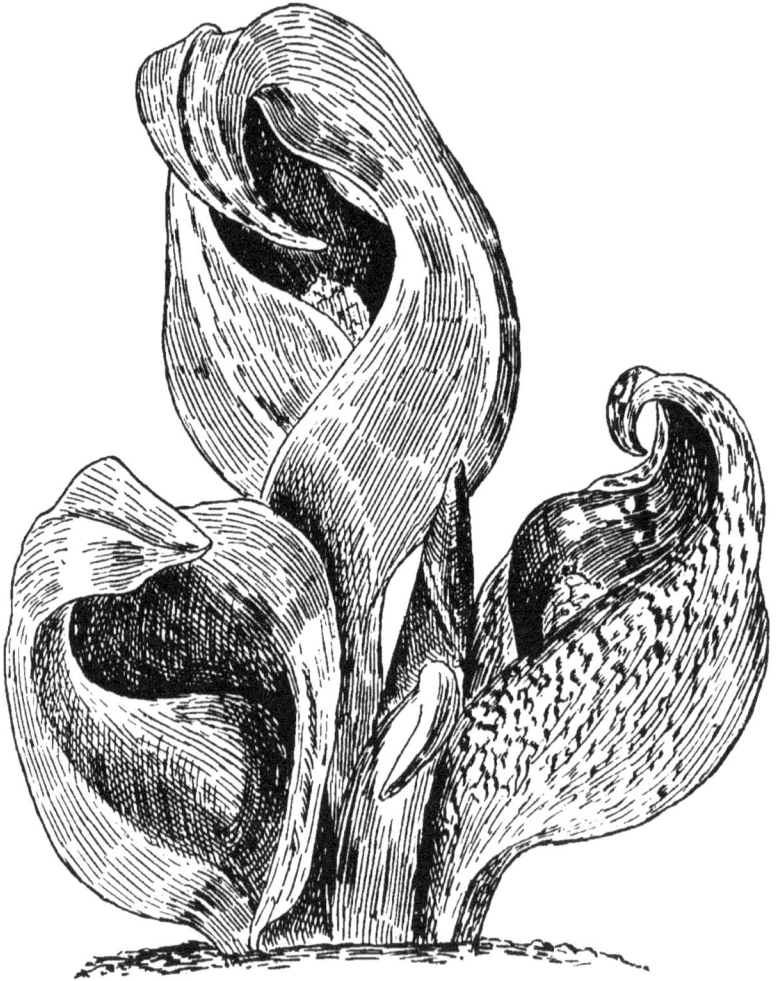

SKUNK CABBAGE.—*S. fœtidus.*

277

object, suggestive of a great snail when it lifts itself fairly above its muddy bed. When one sees it grouped with brother-cabbages it is easy to understand why a nearly allied species, which abounds along the Italian Riviera, should be entitled "Cappucini" by the neighboring peasants, for the bowed, hooded appearance of these plants might easily suggest the cowled Capuchins.

It seems unfortunate that our earliest spring flower (for such it undoubtedly is) should possess so unpleasant an odor as to win for itself the unpoetic title of skunk cabbage. There is also some incongruity in the heading of the great floral procession of the year by the minute hidden blossoms of this plant. That they are enabled to survive the raw March winds which are rampant when they first appear is probably due to the protection afforded them by the leathery leaf or spathe. When the true leaves unfold they mark the wet woods and meadows with bright patches of rich foliage, which with that of the hellebore, flash constantly into sight as we travel through the country in April.

It is interesting to remember that the skunk cabbage is nearly akin to the spotless calla lily, the purple-mottled spathe of the one answering to the snowy petal-like leaf of the other. Meehan tells us that the name bear-weed was given to the plant by the early Swedish settlers in the neighborhood of Philadelphia. It seems that the bears greatly relished this early green, which Meehan remarks "must have been a hot morsel, as the juice is acrid, and is said to possess some narcotic power, while that of the root, when chewed, causes the eyesight to grow dim."

WILD GINGER.

Asarum Canadense. Birthwort Family.

Leaves.—One or two on each plant, kidney or heart-shaped, fuzzy, long-stalked. *Flower.*—Dull purplish-brown, solitary, close to the ground on a short flower-stalk from the fork of the leaves. *Calyx.*—Three-cleft, bell-shaped. *Corolla.*—None. *Stamens.*—Twelve. *Pistil.*—One, with a thick style and six thick, radiating stigmas.

Certain flowers might be grouped under the head of "vegetable cranks." Here would be classed the evening primrose, which only opens at night, the closed gentian, which never opens

PLATE CIII

WILD GINGER.—*A. Canadense.*

279

at all, and the wild ginger, whose odd, unlovely flower seeks protection beneath its long-stemmed fuzzy leaves, and hides its head upon the ground as if unwilling to challenge comparison with its more brilliant brethren. Unless already familiar with this plant there is nothing to tell one when it has reached its flowering season ; and many a wanderer through the rocky woods in early May quite overlooks its shy, shamefaced blossom.

The ginger-like flavor of the rootstock is responsible for its common name. It grows wild in many parts of Europe and is cultivated in England, where at one time it was considered a remedy for headache and deafness.

JACK-IN-THE-PULPIT. INDIAN TURNIP.

Arisæma triphyllum. Arum Family.

Scape.—Terminated by a hood-like leaf or spathe. *Leaves.*—Generally two, each divided into three leaflets. *Flowers.*—Small and inconspicuous, packed about the lower part of the fleshy spike or spadix which is shielded by the spathe. *Fruit.*—A bright scarlet berry which is packed upon the spadix with many others.

These quaint little preachers, ensconced in their delicate pulpits, are well known to all who love the woods in early spring. Sometimes these " pulpits " are of a light green veined with a deeper tint ; again they are stained with purple. This difference in color has been thought to indicate the sex of the flowers within—the males are said to be shielded by the green, the females by the purple, hoods. In the nearly allied cuckoo-pints of England, matters appear to be reversed : these plants are called " Lords and Ladies " by the children, the purple-tinged ones being the " Lords," the light green ones the " Ladies." The generic name, *Arisæma*, signifies *bloody arum*, and refers to the dark purple stains of the spathe. An old legend claims that these were received at the Crucifixion :

> Beneath the cross it grew ;
> And in the vase-like hollow of the leaf,
> Catching from that dread shower of agony
> A few mysterious drops, transmitted thus
> Unto the groves and hills their healing stains,
> A heritage, for storm or vernal shower
> Never to blow away.

PLATE CIV

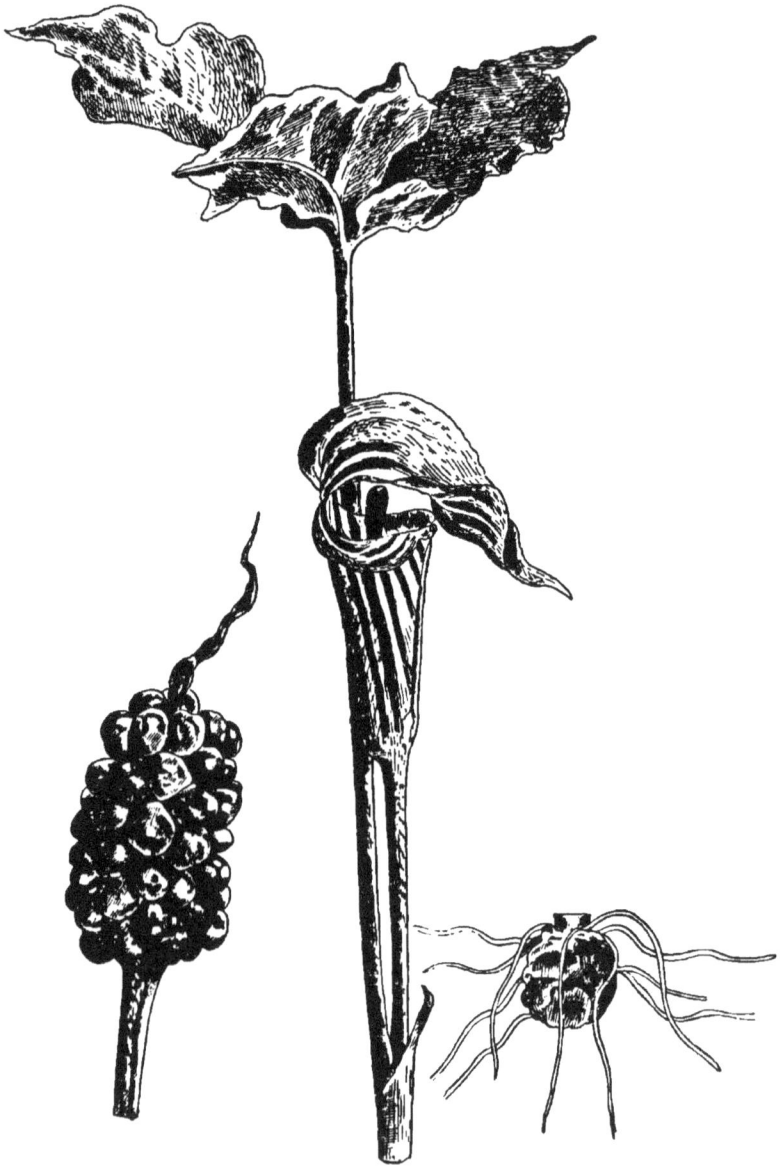

Fruit.

Corm.

JACK-IN-THE-PULPIT.—*A. triphyllum.*

281

The Indians were in the habit of boiling the bright scarlet berries which are so conspicuous in our autumn woods and devouring them with great relish ; they also discovered that the bulb-like base or *corm*, as it is called, lost its acridity on cooking, and made nutritious food, winning for the plant its name of Indian turnip. One of its more local titles is memory-root, which it owes to a favorite school-boy trick of tempting others to bite into the blistering corm with results likely to create a memorable impression.

The English cuckoo-pint yielded a starch which was greatly valued in the time of Elizabethan ruffs, although it proved too blistering to the hands of the washerwomen to remain long in use. Owing to the profusion with which the plant grows in Ireland efforts have been made to utilize it as food in periods of scarcity. By grating the corm into water, and then pouring off the liquid and drying the sediment, it is said that a tasteless, but nutritious, powder can be procured.

ALUM-ROOT.

Heuchera Americana. Saxifrage Family.

Stems.—Two to three feet high, glandular, more or less hairy. *Leaves.* —Heart-shaped, with short, rounded lobes, wavy-toothed, mostly from the root. *Flowers.*—Greenish or purplish, in long narrow clusters. *Calyx.*— Bell-shaped, broad, five-cleft. *Corolla.*—Of five small petals. *Stamens.*— Five. *Pistil.*—One, with two slender styles.

In May the slender clusters of the alum-root are found in the rocky woods.

BLUE COHOSH.

Caulophyllum thalictroides. Barberry Family.

Stems.—One to two and a half feet high. *Leaf.*—Large, divided into many lobed leaflets ; often a smaller one at the base of the flower-cluster. *Flowers.*—Yellowish-green, clustered at the summit of the stem, appearing while the leaf is still small. *Calyx.*—Of six sepals, with three or four small bractlets at base.—*Corolla.*—Of six thick, somewhat kidney-shaped or hooded petals, with short claws. *Stamens.*—Six. *Pistil.*—One. *Fruit.* —Bluish, berry-like.

In the deep rich woods of early spring, especially westward, may be found the clustered flowers and divided leaf of the blue cohosh. The generic name is from two Greek words signifying *stem* and *leaf*, " the stems seeming to form a stalk for the great leaf." (Gray.)

EARLY MEADOW RUE.

Thalictrum dioicum. Crowfoot Family.

One to two feet high. *Leaves.*—Divided into many smooth, lobed, pale, drooping leaflets. *Flowers.*—Purplish and greenish, unisexual. *Calyx.*— Of four or five petal-like sepals. *Corolla.*—None. *Stamens.*—Indefinite in number, with linear yellowish anthers drooping on hair-like filaments (stamens and pistils occurring on different plants). *Pistils.*—Four to fourteen.

The graceful drooping foliage of this plant is perhaps more noticeable than the small flowers which appear in the rocky woods in April or May.

LILY-LEAVED LIPARIS.

Liparis liliifolia. Orchis Family (p. 17).

Scape.—Low, from a solid bulb. *Leaves.*—Two, ovate, smooth. *Flowers.*—Purplish or greenish, with thread-like reflexed petals and a large brown-purplish lip an inch and a half long ; growing in a raceme.

In the moist, rich woods of June we may look for these flowers. The generic name is derived from two Greek words which signify *fat* or *shining*, in reference to " the smooth or unctuous leaves." (Gray.)

BEECHDROPS. CANCER-ROOT.

Epiphegus Virginiana. Broom-rape Family.

Stems.—Slender, fleshy, branching, with small scales ; purplish, yellowish or brownish. *Leaves.*—None. *Flowers.*—Purplish, yellowish or brownish, spiked or racemed, small, of two kinds, the upper sterile, the lower fertile.

These curious-looking plants abound in the shade of beech-trees, drawing nourishment from their roots. The upper open flowers are sterile ; the lower ones, which never expand, accomplish the continuance of their kind.

PINE SAP. FALSE BEECHDROPS.

Monotropa Hypopitys. Heath Family.

A low fleshy herb without green foliage ; tawny, reddish, or whitish. *Flowers.*—Resembling in structure those of the Indian pipe, but clustered in a raceme.

The pine sap is a parasitic plant which is closely allied to the Indian pipe (Pl. XXI.). Its clustered flowers are usually fragrant. The plant is commonly of a somewhat tawny hue, but occasionally one finds a bright red specimen. It flourishes in oak or pine woods from June till August.

RATTLESNAKE-ROOT.

Prenanthes alba.

Height.—Two to four feet. *Leaves.*—The lower cleft or toothed, the uppermost oblong and undivided. *Flower-heads.*—Nodding, composed of white or greenish strap-shaped flowers surrounded by a purplish involucre.

LION'S FOOT. GALL-OF-THE-EARTH.

Prenanthes serpentaria. Composite Family (p. 13).

Height.—About two feet. *Leaves.*—Roughish, the lower lobed, the upper oblong lance-shaped. *Flower-heads.*—Nodding, composed of greenish or cream-colored strap-shaped flowers surrounded by a greenish or purple involucre.

These plants are peculiarly decorative in late summer on account of their graceful, drooping, bell-shaped flower-heads. The flowers themselves almost escape notice, and their color is rather difficult to determine, the purplish or greenish involucre being the plants' conspicuous feature.

The generic name is from the Greek, and signifies *drooping blossom.*

WILD BEAN. GROUND-NUT.

Apios tuberosa. Pulse Family (p. 16).

Stem.—Twining and climbing over bushes. *Leaves.*—Divided into three to seven narrowly oval leaflets. *Flowers.*—Papilionaceous, purplish or chocolate-color, somewhat violet-scented, closely clustered in racemes.

In late summer the dark, rich flowers of the wild bean are found in short, thick clusters among the luxuriant undergrowth and thickets of low ground. The plant is a climber, bearing edible pear-shaped tubers on underground shoots, which give it its generic name signifying *a pear*.

CORAL-ROOT.

Corallorhiza multiflora. Orchis Family (p. 17).

Rootstock.—Much branched, coral-like, toothed. *Stem.*—Nine to eighteen inches high, without green foliage. *Flowers.*—Rather small, dull brownish-purple or yellowish, sometimes mottled with red ; growing in a raceme.

In the dry summer woods one frequently encounters the dull racemes of this rather inconspicuous little plant. It is often found in the immediate neighborhood of the Indian pipe and pine sap. Being, like them, without green foliage, it might be taken for an allied species by the casual observer. This is one of those orchids which are popularly considered unworthy to bear the name, giving rise to so much incredulity or disappointment in the unbotanical.

INDEX TO LATIN NAMES

INDEX TO ENGLISH NAMES

292

INDEX OF TECHNICAL TERMS

.

www.ingramcontent.com/pod-product-compliance
Lightning Source LLC
Chambersburg PA
CBHW021505210326
41599CB00012B/1142